68th Northeastern Weed Science Society Annual Meeting 2014

Proceedings – Volume 68

Philadelphia, Pennsylvania, USA
6 – 9 January 2014

Editor:

Arthur E. Gover

ISBN: 978-1-5108-3339-5

Printed from e-media with permission by:

Curran Associates, Inc.
57 Morehouse Lane
Red Hook, NY 12571

Some format issues inherent in the e-media version may also appear in this print version.

Printed by Curran Associates, Inc. (2017)

Northeastern Weed Science Society (NEWSS)
P.O. Box 25
Woodstown, NJ 08098
USA

Phone: 315-209-7580

northeasternweedscience@hotmail.com

Additional copies of this publication are available from:

Curran Associates, Inc.
57 Morehouse Lane
Red Hook, NY 12571 USA
Phone: 845-758-0400
Fax: 845-758-2633
Email: curran@proceedings.com
Web: www.proceedings.com

Proceedings

of the

Sixty-eighth Annual Meeting

of the

Northeastern Weed Science Society

Arthur E. Gover, Editor

NORTHEASTERN WEED SCIENCE SOCIETY

2014 Sustaining Members

Platinum

BASF The Chemical Company

syngenta

Gold

FMC

DOW

Dow AgroSciences

VALENT

MONSANTO

Silver

BAAR Scientific LLC

ohp Partners with solutions

Gowan USA The Go To Company

DU PONT

UPI

pbi/gordon corporation An Employee-Owned Company

Bronze

Gylling Data Management
TeeJet® Technologies
Weeds, Inc.
Crop Management Strategies
LABSerivices

Table of Contents

Number in parentheses is the title number. Authors and keywords are indexed by title number.

NEWSS EXECUTIVE COMMITTEE - OFFICERS

President	**Dwight Lingenfelter** Penn State University 116 ASI Building University Park, PA, 16802 dxl18@psu.edu
President-Elect	**Gregory Armel** BASF Corporation 26 Davis Drive RTP, NC 27709 gregory.armel@basf.com
Vice President	**Rakesh Chandran** West Virginia University 1076 Agricultural Sciences Building PO Box 6108 Morgantown, WV, 26506 rschandran@mail.wvu.edu
Treasurer	**Erin Hitchner** Syngenta Crop Protection 380 Jefferson Road Elmer, NJ 08318 erin.hitchner@syngenta.com
Past President	**Antonio DiTommaso** Cornell University 903 Bradfield Hall Ithaca, NY 14853 ad97@cornell.edu

NEWSS EXECUTIVE COMMITTEE - MEMBERS

Membership Chair	**Keith Burnell** Syngenta Crop Protection, LLC keith.burnell@syngenta.com
Editor	**Art Gover** Penn State University aeg2@psu.edu
Public Relations	**Todd Mervosh** Connecticut Ag Experiment Station todd.mervosh@ct.gov
Research & Education Coordinator	**Bryan Dillehay** Monsanto Corporation bryan.dillehay@monsanto.com
Sustaining Membership	**John O'Barr** BASF Corporation john.obarr@basf.com
CAST Representative	**Randy Prostak** University of Massachusetts rprostak@umext.umass.edu
Graduate Student Representative	**John Orlowski** University of Kentucky john.orlowski@uky.edu
WSSA Representative	**Jacob Barney** Virginia Tech jnbarney@vt.edu
Science Policy Director	**Lee Van Wychen** Weed Science Society of America lee.vanwychen@wssa.net

NEWSS 2014 MEETING PROGRAM COMMITTEES

Program Committee	Chair:	Greg Armel, BASF
		Rakesh Chandran, West Virginia University
Agronomy	Chair:	Richard Richtmyer, Cornell University
	Chair-elect:	Bryan Dillehay, Monsanto
Ornamentals	Chair:	Kathy Kalmowitz, BASF
	Chair-elect:	Jenny Kao-Kniffin, Cornell
Research Posters	Chair:	Jennifer D'Appollonio-Cote, University of Maine
	Chair-elect:	Tommy Hines, Virginia Tech
Turfgrass and Plant Growth Regulators	Chair:	Aaron Patton, Purdue University
	Chair-elect:	Matthew Elmore, University of Tennessee
Vegetables and Fruit	Chair:	Dan Kunkel, IR-4
	Chair-elect:	Jennifer D'Appollonio-Cote, University of Maine
Vegetation Management and Restoration	Chair:	Jon Johnson, Penn State University
	Chair-elect:	Javier Vargas, University of Tennessee
Weed Biology and Ecology	Chair:	Steven Mirsky, USDA-ARS, Beltsville
	Chair-elect:	Jennifer Ghantous, Univ. of Massachuesetts
Student Paper Contest	Chair:	Toni DiTommaso, Cornell University
		Mark VanGessel, University of Delaware
		Hilary Sandler, University of Massachuesetts
		Dave Yarborough, University of Maine
		Jerry Baron, IR-4
Student Poster Contest	Chair:	Keith Burnell, Syngenta
		Quintin Johnson, University of Delaware
		Don Ganske, DuPont
		Todd Mervosh, CT Ag. Experiment Station
		Art Gover, Penn State University
Photo Judging Contest	Chair:	Darren Lycan, Syngenta
		Anne Marie Pennucci
		Matt Mahoney, Bayer Crop Science
		Javier Vargas, University of Tennessee
		Erin Hitchner, Syngenta

NEWSS COMMITTEES

Site Selection Committee:

Greg Armel	Chair	gregory.armel@basf.com
Dwight Lingenfelter		dxl18@psu.edu
Erin Hitchner		erin.hitchner@syngenta.com
Keith Burnell		keith.burnell@syngenta.com

Awards Committee:

Antonio DiTommaso	Chair	ad97@cornell.edu
Mark VanGessel		mjv@udel.edu
Hilary Sandler		hsandler@umext.umass.edu
David Yarborough		davidy@maine.edu
Jerry Baron		jbaron@aesop.rutgers.edu

Nomination Committee:

Mark VanGessel	Chair	mjv@udel.edu
Sudeep Mathew		samathew@umd.edu
Katherine Ghantous		kghantou@psis.umass.edu
William Curran		wsc2@psu.edu
Larissa Smith		lls14@vt.edu

Resolutions Committee:

Andy Senesac	Chair	afs2@cornell.edu
Courtney Stokes		cas499@cornell.edu
Jennifer D'Appollonio-Cote		jennifer.dappollonio@maine.edu

Collegiate Weed Contest:

Dwight Lingenfelter	Chair	dxl18@psu.edu
Antonio DiTommaso		ad97@cornell.edu
Greg Armel		gregory.armel@basf.com
John Orlowski		john.orlowski@uky.edu

Audit Committee

Erin Hitchner	Treasurer	erin.hitchner@syngenta.com
Melissa Bravo		mbravo@state.pa.us
Todd Mervosh		todd.mervosh@ct.gov
Jim Steffel		jim@labservices.com

NEWSS COMMITTEES (continued)

Archives Committee:

Dan Kunkel	Chair	kunkel@aesop.rutgers.edu
Antonio DiTommaso		ad97@cornell.edu

Weed Science Field Days Committee:

Bryan Dillehay	Chair	bryan.dillehay@monsanto.com
Quintin Johnson		quintin@udel.edu

Past Presidents Committee:

Roy Johnson	Chair	rjoh834880@aol.com
Antonio DiTommaso		ad97@cornell.edu

Herbicide Resistance Plant Committee:

Mark VanGessel	Chair	mjv@udel.edu
Russ Hahn		rrh4@cornell.edu
Bill Curran		wcurran@psu.edu
Brian Olson		bdolson@dowagro.com
Dan Kunkel		kunkel@aesop.rutgers.edu
Dave Mayonado		david.j.mayonado@monsanto.com

Endowment Committee:

Greg Armel	Chair	gregory.armel@basf.com
Renee Keese		renee.keese@basf.com
Jeff Derr		jderr@vt.edu
Erin Hitchner		erin.hitchner@syngenta.com
Hillary Sandler		hsandler@umext.umass.edu
John O'Barr		john.obarr@basf.com

2013 AWARDS, NEWSS/WSSA ANNUAL MEETING

NEWSS Award of Merit: Henry Wilson, Virginia Tech (received by Kurt Vollmer), presented by Mark VanGessel.

NEWSS Fellow Award: Bill Curran, Penn State University

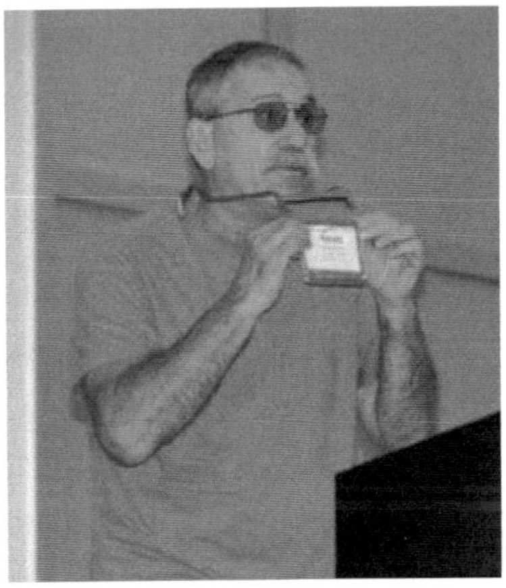

NEWSS Outstanding Researcher: Richard Stalter, St. John's University.

NEWSS Outstanding Researcher: Jim Brosnan, University of Tennesse (received by Matt Elmore).

NEWSS Outstanding Educator: Randy Prostak, University of Massachusetts.

Robert D. Sweet Outstanding Graduate Student Award: Angela Post, Virginia Tech

M. Garry Schnappinger Service Recognition Award: Quintin Johnson, Univ. of Delaware

NEWSS Outstanding Service: Melissa Bravo, (presented by Toni DiTommaso).

2013 AWARDS, NEWSS ANNUAL MEETING (continued)

1st Place, Student Presentation (A.M.):
Kelly Patches, Penn State University

2nd Place, Student Presentation (A.M.):
John Orlowski, University of Kentucky

1st Place, Student Presentation (P.M.):
Katherine Ghantous, University of Massachusetts

2nd Place, Student Presentation (P.M.):
Dan Tekiela, Virginia Tech

1st Place, Student Poster:
Kate Venner, Virginia Tech
(presented by Quintin Johnson)

2nd Place, Student Poster:
Rachel Atwell, North Carolina State University

1st Place, Photo Contest:
Farnaz Kordbacheh, Univ. of Tehran/Cornell University
(presented by Erin Hitchner)

2013 NEWSS COLLEGIATE WEED SCIENCE CONTEST

COLLEGIATE WEED SCIENCE CONTEST

Students from NEWSS and NCWSS
Monsanto Research & Learning Center
Monmouth, Illinois - July 25, 2013

Thanks to all the Monsanto representatives and other volunteers, many who traveled very far, to help make the Weed Contest a success.

NEWSS Contest Winners

North Carolina State University: Graduate Teams
Front (l to r): Sushila Chaudhari, R. Eric Paynter, Alexandra Knight; *Back (l to r)*: Matthew Inman, Nicholas Basinger, Bryan Hicks, Thierry Besancon, Lewis Braswell

NCSU 1st Place Graduate Team:
Thierry Besancon, Sushila Chaudhari, Alexandra Knight, R. Eric Paynter

NCSU 2nd Place Graduate Team:
Nicholas Basinger, Lewis Braswell, Bryan Hicks, Matthew Inman

Penn State 3rd Place Graduate Team:
Kelly Patches, Denyse Schrenker, Clair Keene

1st Place Graduate:
Sushila Chaudhari, NC State University
with Dwight Lingenfelter, NEWSS President

2nd Place Graduate:
Courtney Stokes, Cornell University

3rd Place Graduate:
Clair Keene, Penn State University

2013 NEWSS COLLEGIATE WEED SCIENCE CONTEST (continued)

Univ. of Guelph 1st Place Undergraduate Team:
Jacob Nederend, Jordan Eyamie, Johanna Burrows

1st Place Undergraduate:
Johanna Burrows, Univ. of Guelph
with Dwight Lingenfelter, NEWSS President

Univ. of Guelph 2nd Place Undergraduate Team:
Thomas Heeman, Nicole Bloetjes, Joshua Burrows

2nd Place Undergraduate:
Nicole Bloetjes, Univ. of Guelph

Penn State Univ. 3rd Place Undergraduate Team:
Lyle Stephens, Kaycee Browell,
Phil Sebulsky, Rachel DeFlumeri

3rd Place Undergraduate:
Jordan Eyamie, Univ. of Guelph

CONTRIBUTION OF WEED MANAGEMENT TO BIOENERGY CROP YIELD. L.L. Smith* and J.N. Barney, Virginia Tech, Blacksburg, VA (1)

ABSTRACT

Increasing bioenergy development could help provide a meaningful proportion of our energy needs while creating jobs, decreasing carbon emissions, and stimulating rural economies. There are many options for producing biomass; large-statured perennial grasses have been of particular interest due to their rapid growth, superior biomass, and ability to grow on marginal land. Despite many years of research no single peer reviewed study has evaluated a head-to-head comparison of the yield potential for numerous grass feedstocks under identical growing conditions. In our effort to determine the aboveground yield potential of various bioenergy feedstocks, we evaluated 19 accessions of 9 candidate species over a three-year period in Blacksburg, VA. As weed competition may be an early hindrance to achieving optimal biomass, it was also our objective to evaluate the contribution of weed management to yield in the critical first three years of establishment. We evaluated the following crops: *Andropogon gerardii, Arundo donax*, *Miscanthus* × *giganteus*, *M. sacchariflorus, M. sinensis*, *Panicum virgatum*, *Phalaris arundinacea*, *Saccharum officinarum*, and *Sorghum bicolor.* In year one, overall biomass was low compared to that of year three, and surprisingly, no differences in yield between weedy and weed free plots were observed. In year one, *Sorghum bicolor* (energy sorghum), the only annual species in the study, produced 92% more biomass than the next highest yielding species *M.* × *giganteus.* Despite no measurable differences in year one biomass, the continued lack of a weed management program over consecutive growing seasons appears to negatively influence both establishment and aboveground biomass for most species in this study. Only *P. arundinacea* and two accessions of *M.* × *giganteus* showed no difference in yield between weedy and weed free treatments by the end of the second growing season. In the case of *M.* × *giganteus,* this may be due to its rapid growth and early canopy closure as this species produced the greatest biomass (46,600 kg ha^{-1}), for a perennial species. At the conclusion of the second growing season, *M.* × *giganteus* produced 29% and 76% more biomass than *A. donax* and *P. virgatum* respectively and 5% less than the annual *S. bicolor. M.* × *giganteus* continued to have superior yield in the third growing season in contrast to its parent species *M. sinensis* and *M. sacchariflorus* for which populations have a high level of variability in both height and culm number. Our results suggest that *M.* × *giganteus* may be an ideal candidate for bioenergy production in the Southeastern US and that only minimal weed management may be required.

PHYSIOLOGICAL RESPONSE OF FIVE GOOSEGRASS POPULATIONS TO OXADIAZON. M.C. Cox*, E. Collakova, and S.D. Askew, Virginia Tech, Blacksburg, VA (2)

ABSTRACT

Virginia lies near the northern extreme of preferable growing conditions for bermudagrass (*Cynodon dactylon* L. (Pers.)), and yet over half of Virginia golf courses and a third of athletic fields have bermudagrass turf. In Virginia, cumulative bermudagrass growing degree days can be five times lower than areas further south, winter-kill is an annual threat, and the growing season is approximately 130 days. For these reasons, Virginia turf managers typically do not use root-inhibiting herbicides on bermudagrass turf. Oxadiazon has become the herbicide of choice for preemergence weed control in bermudagrass throughout Virginia, and most practitioners have used it exclusively for over 20 years. Although successful for the previous two decades, oxadiazon at 3.4 to 4.5 kg ai ha^{-1} failed to control goosegrass (*Eleusine indica* L. (Gaertn.)) at a golf course near Richmond, VA. This control failure in 2009 and 2010 prompted studies by Virginia Tech and Bayer CropScience to determine if escaped goosegrass plants had become resistant to oxadiazon. Greenhouse trials conducted in Blacksburg, VA and Clayton, NC evaluated oxadiazon at rates of 0.03 to 4.48 kg ai ha^{-1} and determined that goosegrass plants grown from seed of escaped plants at the aforementioned golf course survived preemergence oxadiazon more than wild-type (WT) plants. No previous occurrences of oxadiazon resistance have been documented anywhere in the world. Several cases of protoporphyrinogen oxidase (PPO) inhibitor resistance have been documented in the last two decades; however, no resistance to oxadiazon has been reported. Additionally, all PPO resistance cases have been from postemergence applications to broadleaf weeds only. Given the uniqueness of potential oxadiazon resistance in goosegrass, research is needed to further characterize the nature of resistance. Laboratory bioassays conducted at the Glade Road Research Facility of Virginia Tech in Blacksburg, VA measured electrolyte leakage from newly germinated goosegrass seedlings as influenced by goosegrass accession and oxadiazon concentration. Subsequent studies measured absorbed and translocated oxadiazon in WT and suspected-resistant (SR) goosegrass seedlings by detecting oxadiazon from plant extracts on a gas chromatography mass spectrometer (GC-MS).

Electrolyte leakage data were fitted to the hyperbolic function, and variance of estimated parameters, *I* and *a,* was analyzed using the general linear model in SAS. Significant difference in *I* and *a* values ($p<0.01$) were noted for the three SR accessions and two WT populations. The WT populations had the highest asymptote for electrolyte leakage percentage, but only WT-1 was significantly higher than SR plants. Electrolyte leakage data suggest that physiological resistance to oxadiazon exhibited by SR plants is not absolute, as the response over oxadiazon rates was similar to WT plants but generally of lower magnitude. Absorption, translocation, and metabolism studies of oxadiazon in the aforementioned goosegrass populations are currently underway and will be discussed at the meeting.

EVALUATING QUALI-PRO PREEMERGENCE LIQUID FORMULATION HERBICIDES FOR CRABGRASS CONTROL AND SELECTIVITY IN BERMUDAGRASS AND PERENNIAL RYEGRASS TURF. A.N. Smith*, S.D. Askew, and J. Corbett, Virginia Tech, Blacksburg, VA (3)

ABSTRACT

Smooth crabgrass (*Digitaria ischaemum* (Schreb.) Schreb. ex Muhl.) is considered one of the most troublesome weeds in turfgrass. There are several preemergence herbicide options available in both granular and liquid formulations. Liquid formulation Quali-Pro (QP) products were compared for efficacy against smooth crabgrass and selectivity in cool- and warm-season turfgrass.

Trials were initiated in fairway height perennial ryegrass (*Lolium perenne* L.) and non-dormant bermudagrass (*Cynodon dactylon* L. (Pers.)) on April 4, 2013. Treatments in bermudagrass were as follows: QP dithiopyr at 280 and 560 g ai ha^{-1}, QP oxadiazon at 3360 g ai ha^{-1}, and QP prodiamine at 1120 g ai ha^{-1} compared to Dimension 2 EW, Ronstar FLO, and Barricade 4 FL at the same rates. Dithiopyr applications at 280 g ai ha^{-1} received a sequential application of the same rate 7 weeks after initial treatment (WAIT). In perennial ryegrass, all treatments were the same except for the exclusion of oxadiazon. A non-treated check was included for comparison. Injury to the turfgrass was rated 1 week after treatment (WAT). Crabgrass control was rated 3, 4, and 5 months after the initial treatment (MAIT).

Oxadiazon injured bermudagrass approximately 40% 1 WAIT, but turf recovered quickly. Dithiopyr injured perennial ryegrass 7% and other treatments did not injure perennial ryegrass. In bermudagrass, crabgrass control was excellent. At 3 MAIT, all treatments controlled crabgrass 96% or better. At 4 and 5 MAIT, applications of dithiopyr at 280 g ai ha^{-1}, QP oxadiazon, QP prodiamine, and Dimension 2 EW at 560 g ai ha^{-1} controlled crabgrass 90% or better. Barricade 4 FL controlled crabgrass 88-92%, and QP dithiopyr at 560 g ai ha^{-1} controlled crabgrass 82%. The best overall control in bermudagrass was dithiopyr at 280 g ai ha^{-1}; both treatments controlled crabgrass 98-100% at 4 MAIT with a reduction to 91-93% control at 5 MAIT. In perennial ryegrass, crabgrass control 3 MAIT was 100% for all treatments. At 4 MAIT, QP prodiamine controlled crabgrass 82%; all other treatments maintained 100% control. At 5 MAIT, QP dithiopyr at 280 g ai ha^{-1} and both rates of Dimension 2 EW controlled crabgrass 90% or better. Prodiamine control decreased to 43-78% control and QP dithiopyr at 560 g ai ha^{-1} controlled crabgrass 86%.

INFLUENCE OF STRAW MULCH ON WEED CONTROL IN CUCURBITS. B. Scott*, M.J. VanGessel, and Q. Johnson, University of Delaware, Georgetown, DE (4)

ABSTRACT

Cereal rye (*Secale cereale* L.) mulch is gaining popularity in no-till cucurbits, particularly pumpkin (*Cucurbita pepo* L.) crops. Its perceived benefits have been publicized to include weed suppression, reduction of fruit disease, and a cleaner pumpkin harvest.

Two trials were conducted in 2012 and 2013 in order to determine 1) if significant weed suppression occurred with rye mulch as compared to conventional tillage and stale seedbed treatments and 2) if an herbicide treatment was still warranted to prevent weed emergence in rye mulch. Each trial was a two-factor factorial design with three replications in 2012 and four replications in 2013. Factors were rye mulch and herbicide treatment.

Treatments in 2012 consisted of: straw mulch at 0, 2500, or 5000 lb/A and treatments were ethalfluralin (Curbit), the premix of ethalfluralin + clomazone (Strategy) or no herbicide. In 2012 all plots were tilled and rye straw was weighed and distributed over plot area. In 2013, plots were either conventionally tilled, or no-till glyphosate application on rye cover and rolled to form a mat. Additionally, a stale seedbed treatment was added in 2013. At the time of rolling, rye biomass was 1700 (low) and 2900 (high) lb/A. In both years, 60 lb/A of nitrogen was dribbled on all treatments prior to vines running.

Palmer amaranth (*Amaranthus palmeri* S. Wats.) and annual grass control ratings were noted both years. In 2013, fruit was sized visually and maturity determined by fruit color. No yield was taken in 2012.

In 2012 at 7 WAT, both Palmer amaranth and annual grass control was reduced by the absence of an herbicide in the low rye mulch. In 2013 at 3 WAT, early season Palmer amaranth control was reduced in all treatments outside the stale seedbed and at yield, late season Palmer amaranth control was reduced in the no-till low rye mulch untreated check and with Strategy as compared to the stale seedbed herbicide treatments. Late season annual grass control was reduced by the absence of an herbicide in the low rye mulch and in the conventional tillage Curbit and untreated check in 2013.

In 2012, pumpkin stunting occurred at the 5,000 lb/A treatment. In 2013, plant stunting in rye mulch was noted and fruit size reduction and delayed fruit maturity occurred in all high rye mulch treatments. This was likely due to nitrogen deficiency.

COMPUTER VISION: A PROMISING TOOL FOR WEED MANAGEMENT. A. Rana* and J. Derr, Virginia Tech, Virginia Beach, VA (5)

ABSTRACT

A common method of weed control in row crops is broadcast applications of postemergence herbicides. Besides covering weed foliage, such applications also apply herbicide to bare soil. If the postemergence herbicide lacks soil residual, then that portion reaching bare ground is wasted. If a system could be developed where the spray is directed only to actively-growing weeds, then reductions in overall herbicide usage would be obtained, especially under low weed densities. A computer vision-based smart spray system has been built at the Hampton Roads Agricultural Research and Extension Center in Virginia Beach. This smart spray system consists of a servo motor-controlled variable cone spray nozzle attached to a pan-tilt system run by servo motors. This smart spray system uses a camera to identify weeds based on color, shape, and size as well as record their respective geo-locations. This image is processed using specially-developed computer vision software. The processed image information is used to twist the spray nozzle on X and Y coordinates to align the spray direction with each weed's respective geo-location. Another servo motor adjusts the cone spray nozzle according to the size of the particular weed of interest so that the spray pattern gives full coverage. A solenoid valve has been used to trigger the amount of herbicide applied. The opening and closing of the solenoid valve is controlled by pulse width modulation signals. This is configured in a way so that larger weeds receive a slightly higher dose of herbicide than smaller weeds. Test results shows that the system is capable of counting target weeds and target weeds one by one using 1.2 to 1.5 ml of herbicide solution. Therefore, this spot spray system is capable of making up to 800 or more individual weed treatments per liter. Since weeds receive a dose based on their size and since the system does not spray bare ground areas, the system should significantly reduce the amount of applied herbicide compared to conventional herbicide applications. This system should also have environmental benefits as less herbicide is applied.

SHADE INDUCED DEMOGRAPHIC VARIATION OF THE INVASIVE PLANT JAPANESE STILTGRASS. D.R. Tekiela* and J.N. Barney, Virginia Tech, Blacksburg, VA (6)

ABSTRACT

Japanese stiltgrass (*Microstegium vimineum* (Trin.) A. Camus var. *imberbe* (Nees) Honda) is an annual C4 grass that has invaded much of the forests in the eastern United States. Unique to stiltgrass is the production of both chasmogamous (CH, open pollinated) and cleistogamous (CL, closed pollinated) seed on the same plant. Though stiltgrass is most successful in high light environments, it is most widely known as a successful invader of forest understories. This success has been attributed to its ability to maintain high fitness even in stressful low light environments. As an annual, the ability to produce seed is critical to stiltgrass; therefore, seedbank dynamics are a critical component to successful invasion.

Our objective was to better understand seedbank demographics of stiltgrass across varying light environments, more specifically, how the seedbank impacts the standing plant population. We established plots along a naturally occurring light gradient in an oak maple hardwood forest with canopy openness ranging from 3 to 43% in southwestern Virginia. We determined that the relative production of CH and CL seed per plant is a function of light availability. More dense populations producing more total seed per plant occurred in high light environment. Additionally, CH seed germination rate was greater (68±4%) than CL seed (27±5%).

Using the data from these studies, we developed a demographic model predicting the steady state of both CH and CL seed in high light and low light populations. Though in all environments there was a greater proportion of CH to CL seed, high light populations had a greater proportion of CH seed relative to CL seed. Regardless of the seedbank, the standing population of stiltgrass was the same across all light environments suggesting stiltgrass is capable of saturating the soil seedbank in high or low light environments. Stiltgrass's ability to saturate the seedbank in understory light conditions may partially explain why this invader has become well established.

DOES NATIVE RANGE SIZE PREDICT INVASIVENESS? A TEST OF THE FEDERAL NOXIOUS WEED LIST. C.A. Ervine*, M. Ho, and J.N. Barney, Virginia Tech, Blacksburg, VA (7)

ABSTRACT

Few studies have been done on the invasiveness of species at a global scale. Results have been conflicting as to whether there are climatic niche shifts in the introduced range. Using Geographic Information Systems, a powerful tool for analyzing spatial data, we assessed the relationship between the size of the native geographic range of invasive plant species and its' successful invasion in introduced ranges. In other words, we are asking the question, does the size and variation of the native range predict the size and success of invasive plant species? Global spatial occurrence data was gathered on all species listed on the US Federal Noxious Weed List (N =112 from the Global Biodiversity Information Facility and from the Early Detection and Distribution Mapping System. Both biotic and abiotic factors determine the success of all species, which translates into spatial geographic extent. Therefore, we used the number of ecoregions each species occurs in to assess the spatial extent of the native and invasive range to test whether the ecological niche had shifted since introduction. We predicted that species with a broader native range, meaning they are present in a greater number of ecoregions, will be more successful invaders than species with smaller native ranges. Assessing whether the extent of the native range is related to successful invasion will provide a better understanding of the invasiveness of species.

MESOTRIONE PLUS TRICLOPYR FORMULATIONS FOR WEED CONTROL IN COOL-SEASON TURF. S.S. Rana*, S.D. Askew, J. Borger, T. Harpster, and D. Lycon, Virginia Polytechnic Institute and State University, Blacksburg, VA (8)

ABSTRACT

Mesotrione (Tenacity™) is a post-emergent, HPPD-inhibiting herbicide that can be used in cool-season turfgrasses to control grassy and some broadleaf weed species. Triclopyr (Turflon® Ester Ultra) may be added to mesotrione to enhance the spectrum of broadleaf weed control and reduce whitening symptoms. Currently, there is no pre-mixed commercial formulation of mesotrione plus triclopyr available to turf managers. Field trials were conducted in Virginia [Virginia Tech (VT), Blacksburg] and Pennsylvania [Syngenta Lawn and Garden (SYN),Pottstown; and Penn State University (PSU), University Park] to determine weed control efficacy of new experimental pre-mixtures compared to Tenacity™ plus Turflon® Ester Ultra as a tank-mix and Q4 Plus (quinclorac+sulfentrazone+2,4-D+dicamba).

The study was initiated on June 1, June 5, and June 19, 2013 at VT, PSU, and SYN, respectively. Treatments included three experimental pre-mixtures each applied at 735 g ai/ha, Tenacity™ at 175 g/ha, Turflon® Ester Ultra at 560 g ae/ha, Tenacity™ plus Turflon® Ester Ultra at the aforementioned rates, and Q4 Plus at 1750 g/ha. A nonionic surfactant at 0.25% v/v was included in all treatments, except Q4 Plus, and a non-treated check was included for comparison. A second application was made six weeks after the initial application. Percent green cover of white clover was assessed at all three locations, smooth crabgrass at VT and SYN, and dandelion at VT and PSU; weed cover ratings were recorded 4 weeks after both applications.

All three experimental pre-mixtures controlled weeds equivalently, and weed control data were pooled for comparison to other treatments. Perennial ryegrass was not injured regardless of herbicide or application timing; however, tall fescue was transiently injured and recovered within 1 week at symptom development. At each respective location, white clover (*Trifolium repens* L.), smooth crabgrass (*Digitaria ischaemum* (Schreb.) Schreb. ex Muhl.), and dandelion (*Taraxacum officinale* G.H. Weber ex Wiggers) cover was significantly less in all treated plots compared to the non-treated check at both 4 and 10 WAIT. In addition, at both 4 and 10 WAIT, experimental pre-mixtures reduced white clover cover (83 and 89%) comparable to Turflon® Ester Ultra (86 and 93%), Tenacity™ plus Turflon® Ester Ultra (89 and 92%), and Q4 Plus (99 and 100%), but greater than Tenacity™ alone (14 and 0% or 50 and 45%). At 4 and 10 WAIT, experimental pre-mixtures reduced smooth crabgrass cover (90 and 94%) comparable to Tenacity™ (91 and 86%), Tenacity™ plus Turflon® Ester Ultra (92 and 96%), and Q4 Plus (92 and 93%), but greater than Turflon® Ester Ultra alone (52 and 42%). All treatments reduced dandelion cover equivalently (greater than 71 and 89%) at 4 and 10 WAIT, respectively. Results from this study indicate that experimental pre-mixtures of mesotrione plus triclopyr controls smooth crabgrass, white clover and dandelion equivalent to Tenacity™ plus Turflon® Ester Ultra and Q4 Plus. Moreover, experimental pre-mixtures controls white clover and smooth crabgrass better than Tenacity™ or Turflon® Ester Ultra applied alone.

PERFORMANCE AND MANAGEMENT OF INTER-SEEDED COVER CROPS IN A TROPICAL SEMI-ARID CROPPING SYSTEM AND THEIR IMPACT ON COTTON YIELDS AND WEEDS. V. Bhaskar*, R. Bellinder, and M.F. Walter, Cornell University, Ithaca, NY (9)

ABSTRACT

This study was conducted to investigate the potential use of inter- seeded cover crops in Vidharba - the cotton belt in India. Summer (February to May) in this region is hot and dry and farming is seldom done during this season, even with the limited irrigation that is available to a small percentage of growers. Weeds do not come up in these extremely dry months, leaving the soil barren and exposed to the sun, causing soil organic carbon reserves to oxidize to unsustainable levels. Current farming practices remove above- ground crop residue after harvest (in January) to use for feed and fuel and this exacerbates the problem. Benefits of cover-cropping are many and fairly well accepted; however, the strategies needed to manage cover crops will influence the outcomes of incorporating them into a cropping system in this region.

Inter-seeding during the cotton (*Gossypium hirsutum* L.) crop was considered as the means of introducing the cover crops because water is scarce after the growing season (June to early January). Two and four trials were conducted in 2012 and 2013, respectively. In both seasons the growth and management of five chosen cover crop species, sesbania (*Sesbania* spp.), gliricidia (*Gliricidia sepium* (Jacq.) Kunth), sorghum x sudan grass hybrid, lablab bean (*Lablab purpureus* L. (Sweet)) and sunnhemp (*Crotalaria juncea* L.) were measured in terms of biomass production and ground cover. The most important cover crop management practice in the trials was to mow them at a height suitable for their regeneration (approximately 18 to 23 cm above ground) when they started to shade the cotton canopy. Cotton, being slow-growing during the first 2 to 3 months of growth, provides an opportunity for the cover crops to establish well enough to withstand the subsequent months of shading.

One trial in both seasons was conducted at the local agricultural research station and the rest were on-farm trials. The data from the 2012 trials showed that the cover crop treatments had no significant effect on the yield of cotton, suppressed weeds, and produced biomass in the range of 9 to 15 tons/ha. It was observed that when the cover crops received an initial 1 to 2 weed-free weeks and germinated well, they reduced weeds to less than $15/m^2$ without any form of mechanical or chemical weed control. As the season neared harvest-time, treatment plots that had been cut back several times tended to be susceptible to weed infestations as the regenerative capacity of the cover crops reduced; but continuing the same operations effectively managed even these plots. Overall, the weed control achieved by the cover crops was more efficient than by the conventional methods employed and more importantly, had no adverse impacts on cotton yields.

IMPACT OF WEED MANAGEMENT SYSTEMS ON GREENHOUSE GAS EMISSIONS.

A.M. Knight*, W.J. Everman, S. Reberg-Horton, and S. Hu, North Carolina State University, Raleigh, NC (10)

ABSTRACT

Worldwide, agriculture accounts for the greatest portion of land with the U.S. having 45% of its land used for agricultural purposes. In addition, agriculture is estimated to contribute greatly to the output of two of the main greenhouse gases suspected of contributing most largely to climate change with methane and nitrous oxide contributing 37 and 59 percent to emissions, respectively. These large percentages are suspected to partially be due to the fact that only one-third of nitrogen applied to cropping systems is actually utilized by the system while the additional two-thirds are lost to the environment. With different agricultural practices contributing differently to these greenhouse gas emissions, finding how different practices contribute to greenhouse gas emissions will help in the recommendation of best management practices for minimal gas emission contributions by agriculture in the southeastern U.S. Field studies were conducted in 2013 at the Center for Agricultural Farming Systems at the Cherry Research Farm in Goldsboro, NC. Long-term plots of conventional no-till, conventional-tillage, conventional crop-hay, organic tillage, organic minimal tillage, and organic crop-hay systems were all used with subplot treatments of weedy and weed-free to measure the flux of the greenhouse gases CO_2, CH_4, and N_2O, 24 to 48 hours after ~1.25 cm or more of rainfall following USDA-ARS GRACEnet Project Protocols. Much greater differences were observed between different systems than between subplots. Minimal differences could likely be attributed to low weed densities in the organic system treatments.

SPRAY VOLUME AND CONCENTRATION EFFECTS ON BROADLEAF WEED CONTROL WITH FE-HEDTA. J. Neal* and C. Harlow, North Carolina State University, Raleigh, NC (11)

ABSTRACT

Fe-HEDTA (trade name: Fiesta) is currently marketed for broadleaf weed control in warm-season and cool-season turf. Application instructions on the label suggest a dilution of 5 oz product per gallon applied in approximately 2.5 to 10 gallons per 1000 sq ft (109 to 436 gallons per acre) spray volume. These guidelines provide a wide range of doses and costs per use. Two experiments were conducted to evaluate the influence of spray carrier volume, spray concentration and dose of Fe-HEDTA on weed control. Fe-HEDTA product was diluted to 1.95%, 3.9% and 7.8% by volume in water and each dilution applied in 50, 100, 200 and 400 GPA spray volume using a CO_2 pressurized sprayer equipped with 8008 flat fan nozzles calibrated to deliver 50 GPA. Higher spray volumes were achieved by multiple (2, 4, or 8) passes over the plot using the same calibrated sprayer. One test was on established *Trifolium repens* and *Taraxacum officinale*. Treatments were applied on October 16, 2012 and again on November 6, 2012. The second site was a low maintenance bermudagrass (*Cynodon dactylon* L. (Pers.)) turf infested with a mixture of winter annual broadleaf weeds including *Veronica persica*, *Veronica hederifolia*, and *Sherardia arvense*. Treatments were applied on March 29, 2013 and re-applied on April 26, 2013. Percent broadleaf weed cover was visually estimated and percent cover estimates were conducted. Both tests were conducted in a randomized complete block design with 4 replications and individual plots were 1 m^2 in size. Data were also averaged by actual Fe-HEDTA dose, across all spray volume and concentrations variables, then plotted against percent control or percent ground cover. Regression equations were fit to these data. Results were consistent between experiments. Fe-HEDTA produced rapid necrosis of broadleaf weed foliage. Foliage turned black within 48 hours of treatment but re-growth was visible within one week of treatment. Weed control increased with increasing concentration and spray volume. However, when control data are plotted against actual Fe-HEDTA dose (across all concentrations and spray volumes), the results fit a logistic model (R^2 values between 0.82 and 0.93) suggesting that the actual dose applied has a greater influence on resulting levels of weed control than do the concentration or spray volume selected.

FURTHER ADVANCEMENTS IN PERENNIAL WEED CONTROL WITH FLAME CULTIVATION. K.M. Ghantous* and H. Sandler, UMass Cranberry Station, East Wareham, MA (12)

ABSTRACT

Interest in reducing chemical inputs into cranberry systems has provided the motivation to evaluate potential nonchemical weed control methods such as flame cultivation (FC). Also known as thermal weeding, FC exposes plants to brief periods of high temperature that causes the water in the plant tissue to expand rapidly, rupturing plant cells and leading to necrosis. Cranberry (*Vaccinium macrocarpon* Ait.) production areas are a continuous canopy of perennial cranberry vines. Ideally FC would be used for spot treatments of small weedy areas, and some cranberry would be incidentally exposed. A series of experiments were designed to study the impact of FC treatments with hand-held propane torches on cranberry, soft rushe (*Juncus effusus* L.), and dewberry (*Rubus* spp.).

The cranberry study showed that vines were able to tolerate FC treatments, which resulted in damage but not mortality. Most perennial weeds in cranberry require repeated treatment for control, regardless of method. Rush was not eliminated by a single FC treatment, however the aboveground biomass was significantly reduced by FC treatments, and treatments were as effective as clipping plants to the ground. Similarly, dewberry was not eradicated from a single exposure to FC but plants showed significant reduction in aboveground biomass a year after treatment, and plants receiving a single August treatment as well as plants that received two treatment (June/July and July/August) also had a significant reduction in root biomass one year after treatment.

Weed control with FC shows promise for small areas within the cranberry canopy, as well as in larger areas where cranberry damage is not a concern, such as ditches and bed edges.

TOLERANCE OF GIANT REED GROWN AS A BIOENERGY CROP TO APPLICATIONS OF POSTEMERGENCE HERBICIDES. B.A. Hicks* and W.J. Everman, North Carolina State University, Raleigh, NC (13)

ABSTRACT

Although many states include giant reed (*Arundo donax* L.) on their noxious weed list, it is being considered as a source of biomass for biofuel production in North Carolina. Since there is little to no information on the effects of herbicides for establishment of giant reed, greenhouse studies were conducted in 2012 and 2013 to observe the tolerance of giant reed to postemergence (POST) herbicides. Our objective was to screen POST herbicides for potential use in the establishment of giant reed as a biomass crop. The study was conducted using greenhouse propagated plants. All plants were established from cuttings into 10.16 cm square pots and allowed to grow to an average height of 25 cm before treatments were applied. Treatments consisted of 2,4D @ 0.475 lb a.i./A, aminopyralid @ 0.109 lb a.i./A + NIS 0.50% v/v, atrazine @ 1 lb a.i./A + COC 1%v/v, bentazon @ 1 lb a.i./A + COC 0.25% v/v, bispyribac-sodium @ 0.0285 lb a.i./A, bromoxynil @ 0.375 lb a.i./A, carfentrazone @ 0.0313 lb a.i./A + COC 1% v/v, chlorsulfuron @ 0.047 lb a.i./A + NIS 0.25%v/v, clethodim @ 0.0682 lb a.i./A + NIS 0.25% v/v, clopyralid @ 0.188 lb a.i./A, cloransulam @ 0.016 lb a.i./A + NIS 0.25% v/v, clorimuron @ 0.012 lb a.i./A + NIS 0.25% v/v, diacamba @ 0.25 lb a.i./A, fluazifop @ 0.141lb a.i./A + NIS 0.25% v/v, fomesafen @ 0.353 lb a.i./A + NIS 0.25% v/v, glufosinate @ 0.402 lb a.i./A + AMS 8.5lbs/100 gal, halosulfuron @ 0.031 lb a.i./A + NIS 0.50% v/v, imazamox @ 0.039 lb a.i./A + NIS 0.25% v/v, imazapic @ 0.063 lb a.i./A + NIS 0.25%v/v, mesotrione @ 0.094 lb a.i./A + COC 1% v/v, metsulfuron methyl @ 0.038 lb a.i./A, nicosulfuron @ 0.031 lb a.i./A + NIS 0.25% v/v, pinoxaden @ 0.054 lb a.i./A, pyraflufen ethyl @ 0.001 lb a.i./A, quinclorac @ 0.75 lb a.i./A + MSO 1.5 pt/A, rimsulfuron + thifensulfuron @ 0.018 lb a.i./A, sulfometuron @ 0.094 lb a.i./A + NIS 0.25% v/v, sulfosulfuron @ 0.031 lb ai ac+ NIS 0.50% v/v, tembotrione @ 0.082 lb ai ac + MSO 1% v/v, tembotrione + thiencarbazone-methyl @ 0.081 lb a.i./A + COC 1% v/v, thifensulfuron @ 0.004 lb a.i./A + NIS 0.25% v/v, topramazone @ 0.016 lb a.i./A + MSO 1% v/v, trifloxsulfuron @ 0.005 lb a.i./A + NIS 0.25% v/v. The greenhouse study resulted in clethodim, fluazifop, sulfometuron, imazapic having greater than thirty percent injury, with clopyralid, tembotrione + thiencarbazone-methyl, and pinoxaden having less than fifteen percent injury. All other treatments fell between these two percentages.

DISTRIBUTION OF HERBICIDE RESISTANT ITALIAN RYEGRASS IN NORTH CAROLINA. Z.R. Taylor* and W.J. Everman, North Carolina State University, Raleigh, NC (14)

ABSTRACT

Italian ryegrass (*Lolium perenne L.* ssp. *multiflorum* (Lam.) Husnot) has become a major weed problem in winter wheat in the southeast, particularly as herbicide resistant biotypes have developed and spread. While we know resistant Italian ryegrass is a problem, little is known about what types of resistance are out there and what areas are affected. In the spring of 2012, a survey was conducted across the major wheat growing areas of North Carolina. Italian ryegrass seed was sampled from fields found in a grid pattern spaced approximately 10 latitude and longitude minutes. Seed samples were taken from 136 field sites. Those seeds were grown in a greenhouse and then treated with postemergence herbicides commonly used for Italian ryegrass control in winter wheat production. The treatments consisted of Powerflex (pyroxsulam) at 3.5 oz/A plus non-ionic surfactant at 0.5% v/v, Hoelon (diclofop-methyl) at 2.66 pt/A, Axial XL (pinoxaden) at 16.4 fl oz/A, and Osprey (Mesosulfuron) at 4.75 oz/A plus methylated seed oil at 1.5 pt/A. We defined resistance as being less than 50% injury, rated visually, after 4 weeks. Initial results indicate resistance to Powerflex in 52% of the tested population, Hoelon in 93% of the tested population, and Axial in 6% of tested populations. Osprey data has not yet been determined and studies are continuing. We found that in all of the locations that showed resistance to Axial, there was also resistance to all other herbicides. This is important to note because it shows that these populations are likely to have resistance to multiple modes of action. Results from this study can help growers determine if resistance is likely in their area, and tailor management strategies to both control and avoid resistant populations.

NOVEL HERBICIDE COMBINATIONS OF HPPD INHIBITORS WITH PLANT GROWTH REGULATORS FOR IMPROVING PALMER AMARANTH CONTROL. R.E. Paynter*, W.J. Everman, J. Burton, and T. Gannon, NCSU, Raleigh, NC (15)

ABSTRACT

One of the most problematic and influential weed species on research in agriculture is Palmer amaranth (*Amaranthus palmeri* S. Wats.). As of late, Palmer amaranth has taken over southern row crop fields and is moving northward. Further exacerbating the Palmer amaranth issue, certain biotypes are resistant to multiple modes of action, therefore new technologies and formulations are needed to properly deal with the problem at hand. The research in this study was conducted by using the HPPD-inhibiting herbicides topramezone, isoxaflutole, or mesotrione applied at 4.63 g ai/ ha, 21.56 g ai/ ha, and 26.36 g ai/ha, respectively, in combination with transport inhibitors diflufenzopyr (DFF), cyclanilide (CYC), or napthalam (NPA). The study was conducted in a greenhouse at North Carolina State University with application applied via a spray chamber. Each herbicide rate evaluated is lower than typical field rates and was tested with a range of concentrations of additives. DFF was added to each herbicide at 2.5 g ai/ ha, 5 g ai/ ha, 7.5 g ai/ ha, or 10 g ai/ ha; while CYC or NPA were added at rates of 52.7 g ai/ ha, 105.42 g ai/ ha, 158 g ai/ha, or 210.9 g ai/ ha). Significant increases in herbicide activity were observed where isoxaflutole was applied with DFF and mesotrione was combined with NPA when compared to the individual herbicide alone. Furthermore, significant effects of fresh weight reductions were observed for topramezone plus DFF, mesotrione plus NPA, Isoxaflutole plus CYC, and topramezone plus CYC when compared to herbicides with no additives. Although further research is needed to refine combinations and optimize formulations this study shows the potential for additives to play an important role in managing resistant Palmer amaranth in the future.

BURNDOWN PROGRAMS FOR TROUBLESOME SOYBEAN WEEDS IN NORTH CAROLINA. L.J. Vincent*, W.J. Everman, R.E. Paynter, Z.R. Taylor, and B.A. Hicks, North Carolina State University, Raleigh, NC (16)

ABSTRACT

The use of a burndown program before planting soybeans is a critical weed management strategy to control emerged weeds and provide a residual effect to suppress weed pressure as soybean seedlings emerge, particularly in Palmer amaranth (*Amaranthus palmeri* S. Wats.) and horseweed (*Conyza canadensis* (L.) Cronq.). Both of these weed species have become particularly tough for producers to control because of their prolific and unique growth habits. Palmer amaranth has become increasingly popular due to its growing abundance across the agricultural landscape, especially in the Southeast. A dioecious *Amaranthus* species, Palmer amaranth adapts quickly, produces hundreds of thousands of seed and has been found to be resistant to several herbicidal modes of action making them one of the toughest weeds to manage. Horseweed is an annual weed, which can have a winter or summer annual life cycle and also possesses a large number of viable seed. In its winter annual cycle, horseweed matures uncharacteristically late into the fall allowing it to compete more aggressively than other winter annuals. Four studies were conducted, two in 2012 and two in 2013 in order to evaluate the level of weed control of several burndown herbicides. Trials were conducted at three locations across North Carolina; Edgecombe County, at the Peanut Belt Research Station near Lewiston-Woodville, and at the Upper Coastal Plain Research Station near Rocky Mount. The list of select herbicides evaluated in the studies includes; 2,4-D, glyphosate, Canopy, Classic, Envive, Express, Leadoff, Sharpen, and Valor at various rates and timings. The aforementioned herbicides were followed with pre-emergence herbicides and post emergence herbicides in a season long weed management program. Ratings of visual weed control were taken periodically throughout the growing season with yield data collected at the end of the season to determine effective weed management of Palmer amaranth and horseweed with various treatment combinations.

EFFECTS OF GROWTH REGULATOR RATE AND APPLICATION TIMING ON SORGHUM GROWTH AND YIELD. T.E. Besancon*, W.J. Everman, R. Weisz, R. Riar, and M.K. Bansal, North Carolina State University, RALEIGH, NC (17)

ABSTRACT

North Carolina growers produced sorghum (*Sorghum bicolor* (L.) Moench ssp. *bicolor*) on 100,000 acres in year 2013. This acreage is expected to increase in the future due to an assured market, grower enthusiasm, low production cost, drought tolerance and a lowered risk when compared to corn, especially on sandy soils. Since sorghum is an old crop with renewed interest, this is the first time it has been grown on a large scale in North Carolina. This crop may be a good fit in the overall cropping system of the state. However, limited information on production and weed management is available in the state. Weed control is an important aspect of profitable crop production which has a bearing on the success of a new crop, and in determining its fit in a cropping system. In order to address the issues of successful weed management under the given environment and cropping system, studies were conducted in 2012 and 2013 at the Upper Coastal Plain Research Station (Rocky Mount, NC) and at the Central Crops Research Station (Clayton, NC). The objective was to evaluate the growth and yield response of sorghum to different rates of 2,4-D amine (100, 217 and 333 g ai.ha^{-1}) and one rate of Dicamba (280 g ai.ha^{-1}) applied post-emergence beyond the recommended growth stage (15-20 cm tall), at 25, 35, 46, 56, 66, and 74 cm height. Significant reduction in yield was observed with later applications of both 2, 4-D and dicamba compared to earlier applications. No difference in yield reduction was observed across the different rates of 2,4D-amine. Yield reduction was significantly more important for dicamba compared to 2,4-D amine. No effect of stage of application was observed on yield reduction for both herbicides.

THE BIOCONTROL AGENT AILANTHUS WILT CAN BE TRANSFERRED TO UNINFECTED TREE-OF-HEAVEN USING RAW INOCULUM. T.A. Rung* and A.E. Gover, Penn State University, University Park, PA (18)

ABSTRACT

Tree-of-heaven (*Ailanthus altissima* (Mill.) Swingle, AILAL) is a persistent weed tree that has proven invasive in nearly all landscapes. The fungal pathogen *Verticillium nonalfalfae* (formerly *Verticillium albo-atrum*) has been identified as a potential biological control for AILAL, as collected isolates have shown virulence, and limited host range. Currently, USDA-APHIS allows movement of this pathogen within state borders where it has been isolated. Biological control may be particularly useful as an alternative control of AILAL as it has shown progressive control within infestations after introduction. This study observed the effects of moving raw inoculum from infected AILAL to un-infected stands, compared to inoculating uninfected stems with a provided purified inoculum, at Canoe Creek State Park, Hollidaysburg, PA. Transfers were conducted June 21 and July 3, 2013. Methods included placing soil from the base of infected stems to the base of uninfected stems where the roots were debarked to expose vascular tissue; making an aqueous slurry of bark, twig, and rachis tissue from infected stems and applying the filtrate to fresh stem cuts on uninfected stems; and transferring stem wedges from infected to uninfected stems. In a single growing season, target trees inoculated via filtrate application or stem wedge were observed to display symptoms of infection, and exhibited significant dieback. Subsequent observations in 2014 will verify if this method provides the same effect observed from application of cultured inoculum. If found to be effective, raw inoculum transfer of *V. nonalfalfae* could prove to be an efficient and effective means of controlling AILAL, as there several known sites in PA that can serve as raw inoculum sources.

POTENTIAL PPO HERBICIDES FOR PROCESSING CUCUMBERS. E. Borgato*, S.A. Mathew, B. Scott, and M.J. VanGessel, University of Delaware, New Wark, DE (19)

ABSTRACT

Pickling cucumbers (*Cucumis sativus* L.) are produced on 5,000 to 7,000 acres in the mid-Atlantic region. All of these acres are machine harvested which requires effective weed control to maximize fruit recovery. Experimental herbicides for potential use in pickling cucumbers need to be evaluated for crop safety and weed control. This project was designed to evaluate PPO-inhibiting herbicides for potential use in pickling cucumbers.

Pickling cucumbers ('Vlaspik' and 'Expedition') were planted July 31 and herbicides were applied immediately after planting. The two pickling cucumbers varieties are the two most commonly planted varieties in the region. PPO herbicides were four rates of sulfentrazone plus carfentrazone (0.07, 0.1, 0.14, and 0.21 lb ai/A) (Spartan Charge), fomesafen at 0.125, 0.187, and 0.25 lbs ai/A, salfufenacil at 0.02 lb ai/A, or flumioxazin at 0.3 lb ai/A. All PPO herbicides were applied with ethalfuralin 0.6 lb ai/A. Ethalfluralin was applied alone at 0.6 and 1.1 lb ai/A.

Main plots were herbicide treatments and varieties were sub-plots. There were three replicates in this trial. Palmer amaranth (*Amaranthus palmeri* S. Wats.) control was similar for all treatments containing a PPO-inhibiting herbicide (83 to 98%). Ethalfluralin at 0.6 and 1.1 lb ai/A provided 63 and 73% control, respectively.

Injury was similar for both varieties and no variety by herbicide interaction was observed. At 2 weeks after planting, ethalfluralin alone was the only treatment with less than 5% injury. Injury from the lowest rates of sulfentrazone or fomesafen or flumioxazin ranged from 14 to 21%. All other treatments were >25% injury. At harvest, only the low rate of ethalfluralin had no stunting observed. Flumioxazin, the high rate of ethalfluralin, and the lowest rate of sulfentrazone were similar (8, 11, and 12%, respectively). All other herbicides were >18%, with sulfufenacil and the two highest rates of sulfentrazone or fomesafen exhibiting >30% stunting.

The highest yield treatments were in plots treated with the lowest rate of sulfentrazone, flumioxazin, and the low rate of ethalfluralin. There was no significant variety by herbicide interaction.

Preliminary results from this trial need further evaluation for consistency of crop safety and weed control. Low rates of some PPO-inhibiting herbicides may have utility in processing cucumber.

ENLIST™ SOYBEAN TOLERANCE TO ENLIST DUO™ HERBICIDE. B.D. Olson*, J.M. Ellis, D.C. Ruen, and E.F. Scherder, Dow AgroSciences LLC, Geneva, NY (20)

ABSTRACT

Previous research with Enlist™ soybean (*Glycine max* (L.) Merr.) across the Mid-South and Midwest, from 2008 through 2012, demonstrated robust tolerance to 2,4-D when applied preemergence or postemergence. In 2012 and 2013, trials were initiated to evaluate injury to Enlist E3™ soybean following applications of Enlist Duo™ herbicide, a proprietary blend of 2,4-D choline and glyphosate, applied at 2185 and 4370 g ae/ha, 2,4-D choline 1065 and 2130 g ae/ha, glyphosate 1120 and 2240 g ae/ha, glufosinate 542 and 1084 g ae/ha and 2,4-D choline + glufosinate at 1065 + 542 and 2130 + 1084 g ae/ha. Single herbicide treatments were applied at V6 and R2 growth stages.

Enlist E3 soybean demonstrated robust tolerance to 2,4-D choline, glyphosate and Enlist Duo across all application timings and rates. Regardless of rate or application timing, injury averaged less that 1% for 2,4-D choline and glyphosate. For Enlist Duo, overall injury with 2815 g ae/ha was 5% or less at any single application timing seven days after treatment. At the 4370 g ae/ha rate, initial injury increased slightly over the 2185 g ae/ha rate yet was negligible by 14 DAT. At seven days after treatment, single applications of glufosinate applied at either V6 or R2, injury averaged 3-4% and 9-10% for 542 and 1084 g ae/ha, respectively. The addition of 2,4-D choline to glufosinate resulted in an 1-2% increase in crop response compared to glufosinate applied alone. Injury observed at seven days after treatment with either glufosinate or 2,4-D choline + glufosinate averaged less than 5% by 14 DAT.

Enlist E3 soybeans are a joint development of Dow AgroSciences and MS Technologies.

ENHANCING STAKEHOLDER INDENTIFICATION SKILLS WITH NEW METHODS OF DISPLAYING INVASIVE WEEDS. A. Senesac*, Cornell Cooperative Extension, Riverhead, NY (21)

ABSTRACT

Identifying weeds early on in their invasion of landscapes and natural areas is a key component of a sound integrated pest management (IPM) plan. Some weeds have become so invasive on Long Island that Suffolk and Nassau Counties passed legislation in 2007 to slow their spread. These 'Do Not Sell' laws ban the sale, introduction, and propagation of 63 invasive, nonnative plants. As part of the Cornell Cooperative Extension mission to help educate Suffolk County residents about their environment, we have constructed a portable display of 40 invasive plants that are listed in the county's 'Do not Sell' Law. The 'weed wagon' features 40 living plants growing in containers arranged in a tiered display. The wagon can be transported during the year to various venues so that interested audiences can casually view and study the live specimens. Each plant is labeled and accompanied by information about the nature of its invasiveness. The goal of the project is to provide the general public a demonstration of 'what to look for' when identifying and removing invasives throughout Suffolk County. Additionally, we are using smaller more portable displays to bring seedling weeds in the off-season to venues that cannot accommodate the larger display. These utilize vertical plastic columns that allow for 30 weed species to be displayed in a single Flower Tower®. These allow for greater ease of transport and instructional facility. These new tools will be employed in both the growing season and other times of the year when interested stakeholders gather.

EVALUATING THE INVASIVE POTENTIAL OF MISCANTHUS BIOFEEDSTOCKS.
D.P. Matlaga*, N. West, and A. Davis, Susquehanna University, Selinsgrove, PA (22)

ABSTRACT

Concern remains over the potential for bioenergy crops to escape cultivation and establish invasive populations. This concern has motivated much speculation but few studies have evaluated both the population trajectory and community impacts of bioenergy crops outside of cultivation. Using both experimental and modeling approaches we study the perennial grass *Miscanthus x giganteus*, a candidate bioenergy crop in the northern US maize belt. To understand the demographic performance of M. x giganteus outside of cultivation we collected data on 13 plantings in unmanaged lands throughout Illinois. We use these demographic data to parameterize an integrodifference equation model in order to identify thresholds for reproductive parameters that will yield a growing population. We found high spikelet production among *M. x giganteus* individuals; however all spikelets were incapable of germination. Our baseline population model, which included no clonal or seed recruitment, yielded a population growth rate slightly < 1 (lambda $= 0.979$), indicating a declining population. According to our modeling results successful recruitment via rhizome fragments must exceed 20% with greater than 1 fragment produced per plant to obtain a growing population. Thresholds to achieve a growing population via viable seed production were considerably lower, implying that efforts to develop fertile lines of this species should be pursued with caution.

THE EFFECTS OF RESOURCE TRANSLOCATION ON THE SUCCESS OF THE CLONAL RECRUITS OF JAPANESE KNOTWEED. A.M. Budsock*, C.M. Gowton, and D.P. Matlaga, Susquehanna University, Nazareth, PA (23)

ABSTRACT

The primary mode of reproduction in Japanese Knotweed (Fallopia japonica) and cause of invasiveness is clonal reproduction via its dense underground rhizome network. Rhizomes can extend laterally 5-6 m from the parent ramet and may be interconnected, making it difficult to distinguish which rhizome fragment belongs to which above ground shoot. Theoretically, two ramets occupying the area could be the same individual, exchanging nutrients, so it is not known how removing neighboring shoots could impact the success of clonal recruits that may depend on nutrients from neighboring shoots. Rhizomes are important storage sinks for carbohydrates and other nutrients, with known temporal and potentially spatial patterns. The translocation of carbon and nitrogen from above-ground shoots to below-ground rhizomes occurs August-September; however, it is not known how this translocation impacts the success of clonal recruits. By removing above-ground shoots, this study aims to measure the success of consequent clonal recruits in a greenhouse experiment. This experiment was conducted at two sites with dense stands of F. japonica along the West Branch of the Susquehanna River in Selinsgrove, Pennsylvania. There were 45 individual focal ramets per site and two treatment factors: times cut (0 and 1) and neighbor inclusion (0 m, 0.5 m, and 1.5 m). Shoots of the focal ramet received the treatment along with neighboring shoots within the 0 m, 0.5 m, or 1.5 m diameter around the focal ramet. We expect that the height of clonal recruits will decrease when above-ground shoots have been removed and when an increased number of neighbors' shoots have also been removed. Understanding resource translocation could further inform management techniques to decrease the severity of invasion.

EXAMINING THE INFLUENCE OF DISTURBANCE ON THE RELATIVE SUCCESS OF SEXUAL AND CLONAL RECRUITMENT OF JAPANESE KNOTWEED . C.M. Gowton*, A.M. Budsock, and D.P. Matlaga, Susquehanna University, Selinsgrove, PA (24)

ABSTRACT

Japanese knotweed (*Fallopia japonica*) is an herbaceous invasive plant in North American and Europe. Knotweed alters riparian habitats by forming monodominant stands that deeply shade the forest floor. Increased shading can decrease seedling recruitment and increase clonal recruitment. Knotweed can recruit new individuals via both sexual reproduction (seed production) and clonally (rhizome and stem fragmentation). Riparian forest invaded by knotweed may experience regular disturbance due to flooding or mowing. It is important to understand how disturbance impacts sexual and asexual recruitment of this riparian invader as it will provide valuable information for land managers wanting to limit the establishment of knotweed. We addressed this topic using an experimental approach to observe recruitment via three reproductive modes in disturbed and un-disturbed plots. To mimic natural disturbance above ground biomass was removed within a 3.5 diameter area. Within undisturbed plots, present vegetation was left intact. Reproductive propagules (rhizome fragments, stem fragments, and seeds) were planted in both disturbed and undisturbed plots at three sites along the Susquehanna River in Snyder County, PA. Both canopy openness and propagule status (presence of sprouting and height) were censused every four weeks for five months. Preliminary results indicate the disturbance treatment increased canopy openness by 15.7%. The disturbance treatment had little effect on rhizome and stem establishment. Regardless of treatment, rhizome fragments have a higher probability of establishing than seeds or stem fragments. Thus, new recruits of knotweed may be more likely to occur via rhizome fragmentation than seed production or stem fragmentation.

KENTUCKY BLUEGRASS RESEEDING INTERVALS FOR AMICARBAZONE. C. Johnston* and P. McCullough, University of Georgia, Griffin, GA (25)

ABSTRACT

Amicarbazone controls annual bluegrass in Kentucky bluegrass, and turf managers may wish to reseed in treated areas following applications. Field experiments were conducted in Griffin, GA to evaluate reseeding intervals of ‘Midnight’ Kentucky bluegrass following amicarbazone applications at three rates and four timings. Amicarbazone (70 WG) was applied to an irrigated tall fescue lawn at 98, 196, or 392 g a.i./ha at 0, 2, 4, or 6 weeks before seeding. An untreated check was included. One week before seeding Kentucky bluegrass, glyphosate was applied to the field to kill existing vegetation. On the day of seeding (October 23, 2012), the field was scalped with a rotary mower and sliced in two directions. Treatments applied on the day of seeding were applied after Kentucky bluegrass was seeded. There was no interaction for application timing and amicarbazone rate on Kentucky bluegrass establishment on five evaluation dates. All amicarbazone rates and timings had similar Kentucky bluegrass cover to the nontreated. Kentucky bluegrass ground cover increased with time, and averaged 83% (±2) cover across all treatments at 113 days after seeding. Results suggest that Kentucky bluegrass may be seeded on the day of amicarbazone applications without reductions in establishment, compared to the nontreated.

CULTURAL METHODS FOR THE SUPPRESSION OF ITALIAN RYEGRASS IN WINTER WHEAT. Z.R. Taylor* and W.J. Everman, North Carolina State University, Raleigh, NC (26)

ABSTRACT

Italian ryegrass (*Lolium multiflorum*) is one of the most problematic weeds in the production of winter wheat in the southeast. As herbicide resistance issues continue to develop and expand throughout the area, our options for control post emergence continue to decline. As a result we are looking at some cultural practices that may help to suppress the problem when combined with a variety of chemical control plans. One method studied was to improve the crops ability to compete by changing row spacing. We compared wheat planted in 7.5 rows with a drill to those planted in approximately 3.75 inch rows. Differences in control between cultural practices were not seen, however this may have been due to poor weed pressure and growing conditions. A second study was conducted to determine the effect of tillage combined with herbicide treatment on Italian ryegrass management. Wheat grown in both tilled and in no-till conditions received the following treatments; non-treated check, Zidua (pyroxasulfone) at 1.25, 1.5, and 3 oz/A pre, Fierce (Flumioxazin and Pyroxasulfone) at 3 oz/A pre, Zidua at 1.25 oz/A and Sharpen (saflufenacil) at 2 fl oz/A pre, Prowl H2O (pendimethalin) at 2 pt/A at spike, Axiom (Flufenacet and Metribuzin) at 8 oz/A at spike, Axiom at 10 oz/A at spike, Zidua at 1.25 oz/A pre fb Osprey (Mesosulfuron) at 4.75 oz/A and non-ionic surfactant at 1qt/100gal post, Zidua at 1.25 oz/A pre fb Powerflex (pyroxsulam) 3.5 oz/A post, Zidua at 1.25 oz/A pre fb Axial XL (pinoxaden) at 16.4 oz/A post, Osprey at 4.75 oz/A and Zidua at 1 oz/A and non-ionic surfactant at 1 qt/100gal post, Powerflex at 3.5 oz/A and Zidua at 1 oz/A post, Axial XL at 16.4 oz/A and Zidua at 1 oz/A post. We found that control was better for all herbicides in tilled conditions compared to no-till. This study is still in progress, and we have yet to determine if this is due to inverting the seed bank or better soil contact by pre-emergence herbicides.

MISCANTHUS X *GIGANTEUS*: SUSTAINABLE BIOENERGY CROP OR INVADER IN WAITING? L.L. Smith* and J.N. Barney, Virginia Tech, Blacksburg, VA (27)

ABSTRACT

The next few years will be crucial in shaping the future of renewable energy sources, especially for large-statured perennial grasses that hold the most promise. Candidate grasses are ideal not only because of their perennial nature, rapid growth and high annual biomass production, but they also require little management and few inputs, have relatively low pest pressure, and are productive on marginal land. However, it is these ideal set of characteristics that has engendered opinions that candidate bioenergy crops are both invasive and benign. Newly developed fertile varieties of *Miscanthus* × *giganteus* may streamline cultivation while reducing labor and establishment costs. Weed risk assessments previously suggest that this sterile species poses 'low risk' for invasive potential. However, the addition of fertile seed, which dramatically increases the propagule number, must be evaluated for invasiveness. In our effort to evaluate the invasive potential of seeded *M.* × *giganteus* in the Southeast, we chose to compare fertile *M.* × *giganteus* against eight grass species, comprising 19 cultivars, in four environments. We selected the eight grass species to create a head-to-head comparison of species introduced for agronomic purposes that are known invasives in the US (positive controls), as well as species that are known not to be invasive (negative controls). This design allows us to make relative comparisons of our candidate crops along a continuum of invasiveness. Our results indicate that in the establishment year *M.* × *giganteus* growth rate and spread was far lower than that that of johnsongrass (positive control) but biomass and stand counts were far greater than difficult to establish grasses such as big blue stem, switchgrass (negative controls), and reed canarygrass (positive control). *Miscanthus* × *giganteus* also failed to exhibit spread potential in the second growing season while johnsongrass, reed canarygrass, and *M. sacchariflorus* (positive controls) showed varying degrees of clonal spread, away from cultivated plots. It was not until the third growing season that fertile *M.* × *giganteus* individuals were observed outside the cultivated field in regions with both high and low plant competition. Our results are one component of a more comprehensive, long-term study to evaluate the invasive potential of *M.* × *giganteus.* Knowledge gained will help prepare for widespread commercialization of these crops, while planning ahead for improved stewardship.

INVESTIGATING GLYPHOSATE EFFICACY FOR CONTROL OF POKEWEED IN PENNSYLVANIA FIELD CROPS. K.M. Patches* and W.S. Curran, The Pennsylvania State University, University Park, PA (28)

ABSTRACT

Common pokeweed (*Phytolacca americana*) is a perennial broadleaf weed with a large persistent taproot that is also capable of abundant seed production. It has become a frequent problem in agronomic crops in Pennsylvania. Traditionally, tillage was used to manage pokeweed; however, the wide-spread adoption of no-till and a decline in the use of soil residual herbicides in soybean may have allowed pokeweed populations to increase in recent years.

Glyphosate is a commonly used herbicide in agricultural systems. In other Penn State research, glyphosate proved to be a key ingredient for the control of common pokeweed, especially in soybean. In order to examine glyphosate efficacy more closely, four individual experiments were conducted in 2012 and 2013 in central Pennsylvania in fields infested with common pokeweed. The experiments focused on nozzle selection, herbicide rate, carrier volume, and application timing.

In most experiments, individual common pokeweed plants were treated with glyphosate at 0.84 kg ae/ha using air induction nozzles at 187 L/ha in mid-June. However, in the nozzle study, air induction nozzles were compared to standard flat fan nozzles. For the rate study, glyphosate was applied at 0.84, 1.26, and 1.68 kg/ha and for the volume study, 94, 187, and 374 L/ha were compared. The results showed that control was similar for the two nozzles, there was a trend for better control with the higher rates, and a trend for reduced control as volume increased.

For the application timing study, individual common pokeweed plants were treated every two to four weeks from mid-May to late September. For each application timing, treated and untreated plants were collected 8 weeks after application (2012 and 2013) and at the end of the season (2013). Applications before mid-June provided less than 70% control. In mid-June, pokeweed plants were 1.2 meters tall; after mid-June, pokeweed plants averaged 1.5 meters tall. Any application at or after mid-June provided at least 93% control. The results from these experiments showed little advantage to increasing glyphosate rate or altering nozzle selection or carrier volume to improve common pokeweed control. Late spring application timings proved less effective than summer-time treatments which should target control after mid June in central Pennsylvania.

INTERACTION OF ETHEPHON AND METHIOZOLIN ON CREEPING BENTGRASS AND ANNUAL BLUEGRASS PUTTING GREENS. K.A. Venner*, S.D. Askew, and S.J. Koo, Virginia Tech, Blacksburg, VA (29)

ABSTRACT

Methiozolin (PoaCure) is a new herbicide developed by the Moghu Research Center in Daejeon, Korea for safe and selective removal of annual bluegrass (*Poa annua*) from creeping bentgrass (*Agrostis stolonifera*) turf. Methiozolin is a member of the isoxazoline class of chemistry and has been shown to inhibit cell wall biosynthesis and tyrosine aminotransferase. Ethephon (Proxy) is a plant growth regulator which is used to temporarily suppress annual bluegrass seedheads. Previous research conducted on two Virginia golf courses indicated ethephon, when combined with methiozolin, may negatively impact efficacy for annual bluegrass control and creeping bentgrass response. Two studies were initiated in 2012 and 2013 to determine whether or not the addition of ethephon to methiozolin programs would influence annual bluegrass control and creeping bentgrass safety. The first trial was initiated on March 13, 2012 at Blacksburg Country Club (BBCC) in Blacksburg, Virginia on a mixed variety creeping bentgrass putting green. Sequential applications were made on April 16. The second trial was initiated on April 12, 2013 at the Virginia Tech Golf Course (VTGC) in Blacksburg, Virginia on a 'C-19 Congressional' creeping bentgrass putting green. Sequential applications were made on May 14. Methiozolin at 500, 1000 and 2000 g ai ha^{-1} was applied alone or with ethephon at 3818 g ai ha^{-1}. A comparison treatment of 1000 g ai ha^{-1} methiozolin + 3818 g ai ha^{-1} ethephon+ 48 g ai ha^{-1} trinexapac-ethyl as well as an untreated check were also included.

Initial annual bluegrass cover on the putting greens ranged from 48 to74%. At 2 WAIT, annual bluegrass control at BBCC was 32% at 2000 g ai ha^{-1} versus 52% at the VTGC. Methiozolin applied at 500 and 1000 g ai ha^{-1} were not significantly different from one another at BBCC. At VTGC, methiozolin applied at 1000 g ai ha^{-1} controlled annual bluegrass more than 500 g ai ha^{-1}. At 6 WAIT, or 2 weeks after the sequential application, annual bluegrass was controlled 89 to 90% with 1000 to 2000 g ai ha^{-1} and 37% with 500 g ai ha^{-1} methiozolin at BBCC. Annual bluegrass at the VTGC was controlled 90, 39 and 8% with 2000, 1000 and 500 g ai ha^{-1} methiozolin, respectively. Ethephon did not affect overall annual bluegrass control at this rating date. At 8 WAIT, addition of ethephon influenced methiozolin applications at the BBCC, but not at the VTGC. At the conclusion of the study, annual bluegrass was nearly completely controlled with all treatments with the exception of methiozolin applied at 500 g ai ha^{-1}. No undesirable turfgrass injury was observed at the VTGC site in 2013 but severe creeping bentgrass injury was observed at BBCC in 2012. At 8 WAIT in 2012, methiozolin applied alone did not significantly injure bentgrass regardless of rate while combinations of ethephon with 1000 or 2000 g ai ha^{-1} methiozolin injured creeping bentgrass 23 and 80%, respectively. Injury occurred at BBCC when the sequential application was applied immediately following an aeration event. This phenomenon was not observed at the VTGC, as aeration did not occur during the course of the study. These data suggest that depending on site conditions and cultural practices, methiozolin and ethephon should not be applied together to avoid potential, undesirable turfgrass injury. Future research will be performed to address the effects aeration has on the interaction between methiozolin and ethephon on golf course putting greens.

WEED MANAGEMENT AND SORGHUM YIELD AS AFFECTED BY POST HERBICIDES. L.J. Vincent*, W.J. Everman, T. Besancon, and R. Riar, North Carolina State University, Raleigh, NC (30)

ABSTRACT

Sorghum is ancient crop with growing requirements favorable to those present in North Carolina. In 2013, North Carolina producers grew it on more than 70,000 acres. This acreage is expected to increase in 2014 due to assured market, grower enthusiasm, low production cost, drought tolerance and lower risk compared to corn, especially in sandy soils. Although there is rejuvenated interest for its use as an animal feed, this is the first time it has been grown on a large scale within the state in recent years. Limited research based production and management information is available for growers to follow in the Southeast. Weed management is an important aspect of profitable crop production which has a bearing on the success of a new crop, and in determining its fit in a cropping system. To address the issues of successful weed management under the given environment and cropping system, two studies were conducted to evaluate the impact of different herbicides and weed management strategies on sorghum growth and yield. The first study evaluated the effect of two rates of Huskie tank mixed with atrazine and Banvel or 2,4-D, and AMS on weed control and yield, at two locations. The treatment combinations evaluated in this study included; Huskie 13 OZ/A + Atrazine 1 PT/A + AMS 1 LB/A, Huskie 16 OZ/A + Atrazine 1 PT/A + AMS 1 LB/A, Huskie 13 OZ/A + Atrazine 1 PT/A + 2,4-D A 4 FL OZ/A, Huskie 13 OZ/A + Atrazine 1 PT/A + Banvel 4 FL OZ/A and Atrazine 1 PT/A + Buctril 1 PT/A. At one location, there were no differences within treatments, but at the second location, the lesser rate of Huskie tank mixed with atrazine and Banvel or 2,4-D produced greater yields than the higher rate of Huskie tank mixed with atrazine and AMS. The second study tested the impact of Huskie mixed with different adjuvants on crop injury and yield. Various adjuvants were added in with a constant rate of 13 OZ/A of Huskie including; NIS 0.25 % V/V, MSO 1 % V/V, COC 1 % V/V, AMS 1 LB/A, AMS 3 LB/A, 28 % N 1 GAL/A, NIS 0.25 % V/V + AMS 1 LB/A, MSO 1 % V/V + AMS 1 LB/A and COC 1 % V/V + AMS 1 LB/A. All treatment combinations caused significant visible injury on sorghum leaves. However, injury was transient and grain yield was similar for all treatments.

KNOW THY ENEMY: ASSESSING POISON IVY PERFORMANCE. E. Benhase*, J.J. Jelesko, and J.N. Barney, Virginia Tech, Blacksburg, VA (31)

ABSTRACT

Poison ivy (*Toxicodendron radicans*) is a native woody liana found in the Eastern and Midwestern United States that causes an infamous skin rash when touched, which is caused by the chemical urushiol. This dermatitis presents health concerns and eradication challenges to land management professionals and homeowners. While the side effects of touching this plant are well known, the growth requirements and metabolic production of urushiol are relatively unknown. By studying urushiol levels in different poison ivy plant tissues and growth habits in high- and low- light environments, we can design more effective management strategies. Drupes were sterilized, germinated, grown, and harvested in sterile conditions, then analyzed using gas spectrometry – mass spectrometry to identify types and relative concentrations of urushiol conformations. To address hypotheses of poison ivy growth in non-sterile environments, three different accessions were also planted at the Kentland Research Facility in Blacksburg, Virginia in a forest understory environment and field environment to study the effects of light and tree structure on growth. It was found that three different types of urushiol are present in varying levels within different poison ivy tissues, and that decreased light levels increased seedling mortality rate. Though preliminary, these results can be used to better predict poison ivy growth, tailor management strategies, and conduct future research on urushiol metabolism and function.

MAPPING THE INVASION CLIFF: PROPAGULE PRESSURE, SPECIES TRAITS, AND HABITAT CHARACTERS. M. Ho* and J.N. Barney, Virginia Tech, Blacksburg, VA (32)

ABSTRACT

The probability of a species to establish and invade in a nonnative habitat is proportional to increased propagule introductions and increased propagule events, called 'propagule pressure'., A mathematical model of introduction success could allow better predictions of invasion success, as well as the contributing factors to propagule success or failure..

One example of such a model is the invasion cliff, which is a mathematical representation of propagule pressure expressed as $Y = (1-P)^n$; where Y is the probability that at least one propagule in a dispersal event will establish, P is the probability of an individual propagule establishing, and n is the number of arriving propagules into an environment. The P variable comprises the interaction between species traits and environmental characteristics, which comprises a multitude of contributing factors.

A field study was designed to evaluate the interactions between habitat characters (eg, competition levels), propagule pressure, and plant traits (eg seed size and mass) as predictors of invasiveness. The study species included the small seeded *Amaranthus retroflexus* and *Digitaria sanguinalis*, and the large seeded *Abutilon theophrasti* and *Urochloa platyphylla* placed in four different competition environments: full competition, no competition, broadleaves-only, and grass-only. Five levels of propagule pressure were introduced to each habitat: 1, 10, 100, 1000, 10000 seeds.

We recorded emergence, height (cm) and establishment. Averaged over all habitats and propagule levels, large seeded species had 1.4 times more seedlings than small seeded species. Velvetleaf had more than twice the amount of seedlings as redroot pigweed, and 3.4 times more than large seeded broadleaf signalgrass. The results suggest that seed size, a species trait, may be a significant determinate of germination rate in the field. When averaged across all four species, we found that emergence (%) was >2-fold higher in competition free environments compared to completely management environments for every propagule level. Our results indicate that biodiversity and species traits may influence the rate of emergence in weedy and invasive species.

TOLERANCE OF SEVERAL LANDSCAPE ORNAMENTALS TO APPLICATIONS OF PENDIMETHALIN AND DIMETHENAMID-P FOR WEED CONTROL. J.J. Vargas*, J.T. Brosnan, G.K. Breeden, and K.M. Kalmowitz, University of Tennessee, Knoxville, TN (33)

ABSTRACT

Research was conducted at The University of Tennessee, East Tennessee Research and Education Center-Plant Sciences Unit (Knoxville, TN) evaluating tolerance and weed control following preemergence (PRE) applications of dimethenamid-P and mixtures of dimethenamid-P + pendimethalin to landscape ornamentals in the field. Ten herbaceous and woody ornamental species were transplanted into a tilled Sequatchie silt loam soil on 21 June 2013. Ornamental plants evaluated included mondo grass (Ophiopogon japonicus), dwarf naupaka (Dwarf naupaka aemula), sweet potato vine (Ipomoea batatas, cv. "Bright Ideas Lime"), St. John's wort (Hypernicum perforatum), daylily (Hemerocallis spp.), red geranium (Pelargonium zonale, cv. "Rocky Mountain Red"), coral bells (Heuchera micrantha, cv. "Palace Purple"), Virginia sweetspire (Itea virginica, cv. "Henry's Garnet Itea"), Aucuba (Aucuba japonica, cv. "Gold Dust Laurel"), and forsythia (Forsythia suspensa). Plant heights at application were as follows: mondo grass (7 to 13 cm), dwarf naupaka (15 to 20 cm), sweet potato vine (20 to 30 cm), St. John's wort (20 to 25 cm), daylily (20 to 30 cm), red geranium (20 to 30 cm), coral bells (35 to 40 cm), Virginia sweetspire (60 to 90 cm), Aucuba (60 to 90 cm), and forsythia (60 to 90 cm). Plants were subjected to five irrigation cycles from an overhead irrigation system during the three-day acclimation period prior to herbicide treatment. Herbicides evaluated included dimethenamid-P (1100, 2200, and 4400 g ai/ha), dimethenamid-P + pendimethalin (1264 + 1676, 2537 + 3363, and 5074 + 6726 g ai/ha), isoxaben + trifluralin at (1120 + 4480 g ai/ha), and indaziflam (67 g ai/ha). The latter two treatments were included as standards for comparison. A non-treated check was included for comparison as well. Granular treatments were applied by hand while liquid formulations were applied using a CO_2 powered sprayer calibrated to deliver 217 L/ha at 310 kPa. Dimethenamid-P treatments were applied over-the-top while indaziflam was applied in a post-directed manner. All herbicides were applied on 24 June 2013 and 60 days after initial treatment (DAIT) on 23 August 2013. Experimental design was a randomized complete block with three replications. Plots contained all ten ornamental species and natural infestations of carpetweed (Mollugo verticillata), large crabgrass (Digitaria sanguinalis), yellow nutsedge (Cyperus esculentus), horsenettle (Solanum carolinense) and barnyardgrass (Echinochloa crus-galli). Ornamental injury was visually evaluated 7, 14, 28, 56, 67, 74, 88, 116 DAIT using a 0 (i.e., lowest) to 100% (i.e., highest) scale relative to the non-treated check. Carpetweed, large crabgrass, yellow nutsedge, horsenettle, and barnyardgrass cover were assessed on the same dates. At 28 DAIT, injury to mondograss, St. John's wort, Virginia sweetspire and forsythia was minimal (0 to 15%) regardless of treatment. Significant injury to dwarf naupaka (23 to 30%) was observed with dimethenamid-P + pendimethalin at 2537 + 3363 and 5074 + 6726 g ai/ha. However, applications at 1264 + 1676 g ai/ha only resulted in 6% injury. A similar pattern was detected among the three rates of dimethenamid-P applied alone as well. Dwarf naupaka was tolerant (< 3% injury) of both trifluralin + isoxaben and indaziflam 28 DAIT. Injury to sweet potato vine measured 1 to 7% 28 DAIT with all treatments except dimethenamid-P + pendimethalin at 2537 + 3363 and 5074 + 6726 g ai/ha (16 to 28% injury). By 28 DAIT, red geranium, daylily, coral bells, and Aucuba were injured significantly (>15%) by all herbicides. By 88 DAIT, only 0 to 10% injury was observed on mondograss, sweet potato vine, St. John's wort, daylily, and forsythia regardless of herbicide. Injury to dwarf naupaka 88 DAIT also measured 0% for all treatments except trifluralin + isoxaben (100% injury), indaziflam (33% injury), and the 2200 g ai/ha rate of dimethenamid-P. Similarly, Aucuba injury measured < 10% with all herbicides except dimethenamid-P + pendimethalin at 5074 + 6726 g ai/ha. Virginia sweetspire was injured 13 to 15% by all rates of dimethenamid-P + pendimethalin but only 3 to 8% with dimethenamid-P alone. Red geranium injury with the 2537 + 3363 and 5074 + 6726 g ai/ha rates of dimethenamid-P + pendimethalin and the 4400 g ai/ha rate of dimethenamid-P measured 17 to 28% 88 DAIT, similar to trifluralin + isoxaben (21%). Comparatively, no red geranium injury was observed with indaziflam 88 DAIT. Coral bells injury with dimethenamid-P + pendimethalin (22 to 42%) was greater than dimethenamid-P alone (2 to 10%). Carpetweed cover in dimethenamid-P + pendimethalin treated plots was <10% 28 DAIT. Similar responses were observed with trifluralin + isoxaben and dimethenamid-P (4400 g ai/ha) alone as well. By 88 DAIT, dimethenamid-P + pendimethalin (5074 + 6726 g ai/ha) reduced carpetweed cover greater than the other rates evaluated. Similar responses were observed among rates of dimethenamid-P alone as well. Reductions in carpetweed cover with dimethenamid-P + pendimethalin (5074 + 6726 g ai/ha) were greater than those following treatment with trifluralin + isoxaben and indaziflam 88 DAIT. While no significant differences in yellow nutsedge cover reduction were detected among dimethenamid-P containing treatments, all applications reduced yellow nutsedge cover greater than trifluralin + isoxaben and indaziflam by 28 and 88 DAIT. Few differences in horsenettle cover were detected between treatments; however, only dimethenamid-P + pendimethalin treatments did not reduce cover compared to the non-treated check. Injury and weed control responses indicate that at the rates evaluated in this study, dimethenamid-P + pendimethalin and dimethenamid-P may be options for managing carpetweed and yellow nutsedge in landscape plantings of mondo grass, St. John's wort and forsythia. Applications to dwarf naupaka, sweet potato vine, Aucuba, and daylily should also be considered given the weed control and transient injury observed. However, applications of these herbicides to coral bells and red geranium should be avoided.

COMBINATIONS OF TOPRAMEZONE AND CLOQUINTOCET-MEXYL AGAINST CREEPING BENTGRASS AND GRASSY WEEDS. M. Elmore*, J.T. Brosnan, G.K. Breeden, J.J. Vargas, and G.R. Armel, University of Tennessee, Knoxville, TN (34)

ABSTRACT

Creeping bentgrass (CBG) (*Agrostis stolonifera*) is the most widely used cool-season turfgrass species on golf course fairways and tees in the United States, but it is tolerant to few postemergence herbicides. Topramezone is a hydroxyphenylpyruvate dioxygenase-inhibiting herbicide recently registered for use in turfgrass; however, it is not labeled for use on CBG. Previous experiments demonstrated that the safener cloquintocet-mexyl (cloquintocet) applied at 185 g ha^{-1} increased CBG tolerance to topramezone at 37 g ha^{-1}. Data describing CBG tolerance and weed control with topramezone and lower rates of cloquintocet-mexyl are warranted.

Greenhouse experiments were conducted in 2012 and 2013 at the University of Tennessee to evaluate CBG tolerance to topramezone applied with various rates of cloquintocet. Topramezone (37 g ha^{-1}) was applied to mature 'Penncross' CBG alone or in combination with cloquintocet at 185, 93, 37, 28, 19 or 9 g ha^{-1}. All treatments were applied with NIS at 0.25% v/v. Treatments were applied with a water carrier at 221 L ha^{-1} using a spray chamber to mature CBG grown in 6 cm cone-tainers filled with a peat moss, perlite, and vermiculite growing medium. Plants were maintained in a greenhouse under ambient light and fertilized every 10 days with 10 kg N ha^{-1} using a complete (20-20-20) fertilizer. Treatments were in arranged in a completely randomized design with four replications and repeated in time. Injury was evaluated visually on a 0 (no control) to 100% (complete control) and using digital image analysis 7, 14 and 21 days after treatment (DAT). Aboveground biomass was evaluated 21 DAT.

Topramezone in combination with cloquintocet at ≥ 28 g ha^{-1} injured CBG similarly 14 and 21 DAT. For example, cloquintocet at 28 g ha^{-1} reduced CBG injury from 23 to 10% at 14 DAT. Additional greenhouse research was conducted to evaluate weed control from topramezone applied in combination with cloquintocet, and the herbicide metabolism inhibitor malathion. Topramezone was applied at 0, 2, 4, 8, 16, 32, 64 and 128 g ha^{-1} alone and in combination with cloquintocet (28 g ha^{-1}) or malathion (1000 g ha^{-1}) to determine the rate of topramezone required to provide 50% control (I_{50}) of crabgrass (*Digitaria sanguinalis*) and goosegrass (*Eleusine indica*). Treatments were applied to CBG, large crabgrass goosegrass using the same methodology as described previously. Cloquintocet did not increase topramezone I_{50} values for goosegrass or crabgrass which suggest that cloquintocet does not antagonize topramezone efficacy against crabgrass and goosegrass. Malathion did not reduce topramezone I_{50} values for goosegrass or crabgrass.

CBG tolerance to topramezone may be related to metabolism. Future experiments will investigate mechanisms by which cloquintocet and other herbicide metabolism inhibitors affect CBG tolerance to topramezone. Additional research will also evaluate CBG tolerance and weed control efficacy with topramezone-cloquintocet combinations in the field.

NITROGEN FATE IN CORN AND WEEDS AS IMPACTED BY DIFFERENT NITROGEN SOURCES, RATES, AND WEED REMOVAL HEIGHT. A.M. Knight*, W.J. Everman, D. Jordan, R. Heiniger, and T.J. Smyth, North Carolina State University, Raleigh, NC (35)

ABSTRACT

Two of the greatest factors, following genetics, impacting production and yield in agronomic crops are fertility and weed management. The uptake efficiency of nitrogen is dependent upon many factors including tillage system, soil type, crop, weeds, and the amount and type of nitrogen fertilizer applied. The relationship and interaction between crops and weeds is important, and determining how North Carolina corn production may be impacted by different fertilizers could improve nitrogen use efficiency and overall corn yields. Field studies were conducted in 2011 and 2012 at the Upper Coastal Plains Research Station near Rocky Mount, NC and at the Central Crops Research Station in Clayton, NC. Treatment factors included N source, N rate, and weed removal time with a factorial treatment arrangement. The N sources included urea ammonium nitrate (UAN), chicken litter (CCL) and sulfur coated urea (SCU) with rates of 0 kg N/A, 68 kg N/ha, 135 kg N/ha, and 202 kg N/ha. Weed removal times were at 0 (weed free), 8, and 16 cm heights. Significant year, nitrogen source, and weed removal height effects were observed for corn yield. Differences based on year are not surprising considering the differences in weather patterns between the two seasons. Significance based on source could also have been predicted due to the different sources being used with an organic source, and two synthetic sources one of which was a time release fertilizer. When weeds were allowed to remain in the field with corn, the weeds were able to compete with the corn for nitrogen over a greater time period therefore reducing corn yield potential which showed the importance of the critical period of weed removal.

COMPARING FORMULATIONS AND TIMINGS OF DITHIOPYR PLUS FLORASULAM FOR CRABGRASS AND BROADLEAF WEED CONTROL IN COOL-SEASON TURF.

S.S. Rana*, M.C. Cox, D. Loughner, and S.D. Askew, Virginia Polytechnic Institute and State University, Blacksburg, VA (36)

ABSTRACT

Defendor™ (florasulam) is a new broadleaf herbicide from Dow® AgroSciences that is currently available only as a liquid formulation. Dithiopyr is a leading preemergence crabgrass herbicide in the transition zone and northern U.S., and is available in several formulations. The liquid formulation of dithiopyr, Dimension® 2EW, is sold as a co-pack with Defendor™. Since much of the lawn care industry apply dithiopyr as a granule (dithiopyr G), a granular formulation of florasulam (florasulam G) would be desirable. A granular formulation may contain only florasulam G or a combination of florasulam G and dithiopyr G. Application timing is one way to maximize product performance. Therefore, a field trial was conducted to determine if fall applications of florasulam G (on 35-0-4 fertilizer) alone or fall applications followed by sequential application of dithiopyr G plus florasulam G (on 35-0-4 fertilizer) in early spring (before dandelion blooms) can provide weed control and/or bloom suppression compared to sprayable products and industry standard broadleaf herbicides. The study was initiated on a perennial ryegrass fairway on September 20, 2012. Experimental design was a randomized complete block with twelve treatments and four replications. Treatments were arranged in a factorial design with five levels of initial florasulam G at 18.3 g ai/ha [none, September (A), October (B), November (C), and December(D)] and two levels of spring sequential (E) (florasulam G plus dithiopyr G at 259 g ai/ha or none). Comparison treatments included Dimension® 2EW at 426 g ai/ha plus Defendor™ at 15 g ai/ha and Dimension® 2EW plus Cool Power (dicamba+MCPA+triclopyr) at 1.77 kg ae/ha, both applied at application timing E, which was in March. Percent visual cover, control, and bloom reduction ratings were taken for perennial ryegrass, white clover, and dandelion, where applicable, at 0, 3, 4, and 7 weeks after application E (WAE). Smooth crabgrass could not be rated as it was not present in sufficient density for assessment throughout the course of the study. None of the herbicide programs injured perennial ryegrass. At 4 WAE, all florasulam G-containing treatments (with and without sequential) controlled dandelion less (≤50%) than Dimension® 2EW plus Defendor™ (90%) and Dimension® 2EW plus Cool Power (93%), but equivalent to each other. However, florasulam at the application timing A, A+E, and B+E suppressed dandelion bloom equivalent to Dimension® 2EW plus Defendor™ and Dimension® 2EW plus Cool Power. At 7 WAE, Dimension® 2EW plus Defendor™ and Dimension® 2EW plus Cool Power controlled dandelion 100 and 90%, respectively, which was higher than florasulam G-containing treatments (≤39%), both with or without sequential. All treatments, except florasulam G at the D (41%) and B (55%) timings, controlled white clover equivalently (78-100%), 7 WAE. Results from this study indicate that florasulam G applied no later than September (without sequential) or October (with sequential) will likely suppress dandelion bloom equivalent to Dimension® 2EW plus Defendor™ and Dimension® 2EW plus Cool Power® applied only in the spring. However, florasulam G does not seem to control dandelion equivalent to Dimension® 2EW plus Defendor™ or Dimension® 2EW plus Cool Power® through late spring, but does effectively control white clover.

SWALLOWWORT DISPERSAL: SEED TRACKING UNDER FIELD CONDITIONS. C.A. Stokes*, A. DiTommaso, L.R. Milbrath, C.L. Mohler, and T. Whitlow, Cornell University, Ithaca, NY (37)

ABSTRACT

Vincetoxicum rossicum (pale swallowwort) and *V. nigrum* (black swallowwort) are invasive, perennial vines found in natural areas of the northeastern United States and southeastern Canada. Seeds of both species are primarily wind-dispersed in the form of achenes with an attached coma. For this study, "seed" will refer to the complete achene-coma unit. To better characterize the dispersal of these species by wind, a tracking study was initiated in October 2013. Research took place in hay fields near Aurora and Ithaca, NY. Seeds were placed individually in a clothespin attached to the end of a 1.27-cm-diameter PVC pole. A string was attached to the clothespin and used to allow researchers to open the clothespin while crouched down approximately 2 m from the pole, in an effort to minimize wind disturbance. Release heights of 0.75 and 2 m were used. Wind speed was recorded at the moment of release for each seed using a hand-held anemometer, and distance traveled was measured using a tape affixed to the base of the PVC pole. Seeds were recovered and returned to the lab, where seed weight and settling rate were recorded. Settling rate was obtained by dropping individual seeds down an 8-cm-diameter clear plastic tube 1.22 m long. Falling time in seconds was converted to settling velocity in meters per second. Tracking studies took place a minimum of three times at each location, and seed shadows were generated for each release event. Not all seeds were recovered; 4% of the total seeds released were observed traveling too high or far for researchers to follow. Of the recovered seeds, distance traveled ranged from 0.14 to 86.73 m. It is expected that release height, wind speed, settling rate and seed weight will affect distance traveled; data are being analyzed to test this hypothesis. Additionally, models will be tested for fit. Potential models could include exponential, inverse power, or Clark's 2Dt functions.

CORN, SOYBEAN, AND WHEAT PERFORMANCE IN AN ORGANIC ROTATIONAL NO-TILL SYSTEM IN PENNSYLVANIA. C.L. Keene*, W. Curran, and M. Dempsey, Penn State, State College, PA (38)

ABSTRACT

The economic and environmental benefits to be gained from reducing tillage in organic agronomic crops are substantial; however, it is currently difficult to achieve consistent cash crop yields and maintain adequate weed control in reduced tillage organic crop rotations. A three-year, organic rotational no-till corn silage-soybean-wheat rotation was initiated at the Penn State agronomy research farm in central Pennsylvania in 2011. Hairy vetch-triticale and cereal rye cover crops were seeded before corn and soybean, respectively, each year. Rotational no-till was accomplished by terminating the hairy vetch-triticale and rye cover crops with a roller-crimper and no-till planting corn and soybean into the respective cover crop mulches. Moldboard plowing occurred once per year after cash crop harvest. Corn and soybean planting date, variety maturity, and supplemental weed control were tested as independent variables. Early, intermediate, and late planting dates were used to determine the effectiveness of delayed cash crop planting as a cultural weed control strategy. High residue cultivation was used as a supplemental weed control tactic in half of the corn and soybean plots. Short versus intermediate corn and soybean varieties were compared at each planting date.

From 2011 through 2013, corn silage yields ranged from 20,063-44,017 kg ha^{-1}. Soybean grain yields were more similar across years ranging from a low of 1,850 to a high of 4,106 kg ha^{-1}. Wheat grain yield ranged from 2,534 to 3,701 kg ha^{-1}. The cover crop mulches successfully suppressed weeds in most corn and soybean treatments in all years. Background weed biomass in corn ranged from 0-255.8 g m^{-2} in 2011 and 2012. Weed biomass in soybean ranged from 0-134.2 g m^{-2} in 2011 and 2012.

Corn silage yields were highest for the longest maturity cultivar planted at the early planting date. Silage yield of the standard maturity cultivar did not decrease with delayed planting. Soybean grain yields were generally lower at the late planting date but were similar at the early and mid planting dates. The impact of high residue cultivation on corn silage and soybean grain yield was not consistent across years. Wheat yields varied mostly due to the presence of volunteer cereal rye that had successfully produced seed in the previous soybean crop. Despite good weed control, corn silage yields were generally low. Soybean yields were fairly consistent between the early and mid planting dates. The late planting date exhibited the lowest and most variable yields for corn silage and soybean. The loss of wheat yield to volunteer cover crops in the second and third years of the study emphasize the importance of capturing legacy effects in organic cropping systems research.

MISCANTHUS X GIGANTEUS TOLERANCE TO PRE AND POST HERBICIDE APPLICATIONS. B.A. Hicks*, W.J. Everman, and T. Hamblin, North Carolina State University, Raleigh, NC (39)

ABSTRACT

The cultivation of *Miscanthus x giganteus* as a cellulose feedstock for bioenergy production has created a demand for exploration of preemergence (PRE) and postemergence (POST) herbicides that can reduce or eliminate the impact of weed competition. Weed control is particularly important in the establishment years of *Miscanthus x giganteus*. With this in mind, field and greenhouse studies were conducted in 2013 to assess the tolerance of *Miscanthus x giganteus* to PRE and POST herbicides that have activity on grass and broadleaf species. Field studies using POST herbicides were conducted at three different field locations in Clinton, Magnolia and Mt. Olive, North Carolina during 2013. The experiment was conducted using treatments from the following families of chemistry aryloxyphenoxy-porpionate, benzoyl pyrazole, cyclohexanedione, diphenylether, imidazolinone, phenylpyrazole, phosphinic acid, pyrimidinyloxybenzoic acid, quinoline carboxylic acid, sulfonylurea, triazine, triazolinone, triazolone, and triketone. *Miscanthus x giganteus* was rated for injury on a 0 to 100% scale at 7, 14, 21, 28 and 56 days after application. Injury ratings at the 56 day rating ranged from 6 to 79% for products with grass activity and 2 to 12% for products with broadleaf activity. Greenhouse studies were conducted in Raleigh NC using rhizome pieces planted into 10 cm square pots with PRE herbicide applications. The study was conducted with treatments from benzoic acid, chloroacetamide, dinitroaniline, imadazolinone, isoxazole, isoxazolidinone, n-phenylphthalimide, triazine, triazione, triazolinone, triazolone, and triketone chemical families. Plants were rated for injury on a 0 to 100% scale at 7, 14, 21 and 28 days after application with harvest of above and below ground biomass 28 days after emergence. Visible injury ranged from 0 to 62% for all treatments.

INVESTIGATING SYNTHETIC AUXIN COMBINATIONS WITH PLANT GROWTH REGULATORS TO ENHANCE CONTROL OF PALMER AMARANTH. R.E. Paynter*, W.J. Everman, J. Burton, and T. Gannon, NCSU, Raleigh, NC (40)

ABSTRACT

Palmer amaranth (*Amaranthus palmeri*) is a weed species that is found in almost all row crops in the southern United States farmland and is one of the driving forces in research for weed science. Due to resistance across many herbicides/ modes of action (for example glyphosate and ALS) in Palmer amaranth, new technologies and formulations are needed to properly deal with the problem at hand that many growers struggle with on a daily basis. The research in this study was conducted by using auxin growth regulator herbicides dicamba at 69.5 g ai/ ha and 2,4- D at 55.6 g ai/ ha in combination with transport inhibitors diflufenzopyr (DFF), cyclanilide (CYC), or napthalam (NPA) at various rates. The study was conducted in a greenhouse at North Carolina State University and repeated in time. The herbicides were applied using a spray chamber. Additive rates applied with each herbicide were: DFF at 2.5 g ai/ ha, 5 g ai/ ha, 7.5 g ai/ ha, or 10 g ai/ ha; CYC or NPA at 52.7 g ai/ ha, 105.42 g ai/ ha, 158 g ai/ha, or 210.9 g ai/ ha. A significant increase in herbicide activity was observed, when compared to the individual herbicide applied alone for both dicamba and 2,4-D when combined with DFF at 5 g ai/ ha and 7.5 g ai/ ha. Also, a significant increase in activity was observed in dicamba plus cyclanilide at 210.9 g ai/ha 24 hours after treatment and dicamba plus napthalam at 52.7 g ai/ ha or 105.42 g ai/ ha when compared to dicamba applied alone. Fresh and dry weights were recorded at 7 days after treatment. Significant differences in fresh weight and dry weight were noted for the dicamba plus napthalam compared to herbicides with no additives. These combinations have in the potential to expand the tools necessary to combat the growing herbicide resistance problem.

ECOLOGICAL IMPACTS OF EXOTIC PLANTS: ARE TWO INVADERS WORSE THAN ONE? D.R. Tekiela* and J.N. Barney, Virginia Tech, Blacksburg, VA (41)

ABSTRACT

Microstegium vimineum (Japanese stiltgrass) is an annual C4 grass that was introduced to the United States >100 years ago, and is negatively impacting forest understory health through changes in nutrient cycling and forest successional patterns. More recently, *Oplismenus hirtellus spp. undulatifolius* (Wavyleaf basketgrass), a creeping perennial, was introduced into the Baltimore, Maryland area where *M. vimineum* is already well established. Thought to be introduced in 1996, *O. hirtellus* prefers the same low light understory conditions that *M. vimineum* has already invaded. Invaded forests in this area now commonly exist as a mosaic of both species covering the forest floor. Because of the short residence time, very little is known about the biology and ecology of *O. hirtellus.*

Understanding the impact of invasive species is the last but arguably most important step in the invasion process. Though the impacts of *M. vimineum* have been well studied, the impacts of *O. hirtellus* are relatively unknown. Because these species co-exist and *O. hirtellus* it is important that the impacts of this new invader be better understood. Additionally, because *M. vimineum* and *O. hirtellus* exist in intermixed populations, it is important to determine if the presence of both species may produce synergistic or antagonistic ecological impacts.

The site was established in Soldiers Delight Natural Environment Area adjacent to Patapsco State Park, believed to be the initial point of introduction. By removing one or both species and identifying appropriate uninvaded control sites, we have determined the impact of *M. vimineum* and *O. hirtellus* on the resident plant community. Using the Shannon-Wiener index to determine biodiversity, we concluded both *M. vimineum* (H' = 1.68) and *O. hirtellus* (H' = 0.56) reduced plant community biodiversity when compared to the removal treatment (H' = 2.19) but *O. hirtellus* did so to a greater extent. In addition, richness was only negatively impacted by the presence of *O. hirtellus*. This study suggests the newly introduced *O. hirtellus* is the primary driver of plant community composition within intermixed populations of these two invasive species.

THE EFFECT OF LACTOFEN TIMING ON SOYBEAN YIELD AND PHYSIOLOGY. J.M. Orlowski* and C.D. Lee, University of Kentucky, Lexington, KY (42)

ABSTRACT

Recent historically high soybean prices have allowed growers to consider intensive management strategies to maximize yields and profits. These include herbicide applications to the crop for perceived benefits other than weed control. For example, it has been suggested that protoporphyrinogen oxidase (PPO) inhibitor herbicides applied to young soybeans can directly increase yields. It is hypothesized that these herbicides damage the soybean apical meristem stimulating increased lateral branching thereby increasing the number of reproductive nodes and potential yield. However, there is limited published evidence supporting increased lateral branching caused by PPO herbicides. To test the hypothesis of increased lateral branching and increased soybean yield following PPO herbicide applications, a small plot-study was initiated in 2013 at three locations in Kentucky. The study was arranged in a randomized complete block design and maintained weed free. Herbicides (lactofen at 240 g ai ha^{-1} or fomesafen at 600 grams ai ha^{-1}) were applied at the V1, V2, V3 and V4 soybean growth stages. Treatments for comparison were an untreated control, a treatment where the soybean apical meristems were manually removed, and a treatment where all the leaves were removed. These latter two treatments were performed at the same times as herbicide applications. Data collected at harvest included branches $plant^{-1}$, pods $plant^{-1}$, pods $branch^{-1}$, seeds m^{-2}, seed mass, and seed yield. Removal of the apical meristem increased branches $plant^{-1}$ but did not increase seed yield when compared to the untreated control. Herbicide treatments did not change branches $plant^{-1}$. These results indicate that any yield increase observed following applications of PPO herbicides to early vegetative soybeans is not be due to increased lateral branching but is the result of another, as yet unknown, mechanism.

COMPARISON OF WEED SUPPRESSION IN MONOCULTURES AND GRASS-LEGUME MIXTURES OF SUMMER ANNUAL FORAGES. K.A. Bybee-Finley*, Cornell University, Ithaca, NY (43)

ABSTRACT

Summer annual forage crops can be an important source of livestock feed in emergency situations when corn silage and alfalfa crops are compromised or become unavailable. Functionally diverse forage crop mixtures can increase resource-use efficiency and provide weed suppression over forage crop monocultures, especially in low-input cropping systems. The purpose of this research was to explore resource partitioning and facilitation in summer annual forage crops grown mid-summer in the northeastern United States. Four species with different statures and nitrogen acquisition traits: 1) sorghum sudangrass (*Sorghum bicolor* x *S. sudanense,* grass, tall); 2) pearl millet (*Pennisetum glaucum*, grass, short); 3) sunn hemp (*Crotolaria juncea*, legume, tall); and 4) cowpea (*Vigna unguiculata*, legume, short) were seeded at multiple rates in monocultures and in all possible combinations of 2, 3, and 4 species mixtures. A portion of the treatments, including monocultures and the 3 and 4 species polycultures was compared in a separate complementary experiment that was conducted at three different field sites that varied in planting date, climate, and soil type. Low-input conditions were simulated with a reduced seeding rate and by applying 40 kg/ha of total nitrogen in the form of either poultry litter (5-4-3) or mineral fertilizer (10-20-20) prior to seeding crops. Although this is potentially more nitrogen than would be applied prior to legume crops, it represents less than half of the recommended rate for sorghum sudangrass. Aboveground plant biomass was sampled by clipping vegetation in 0.5 m^2 quadrats at 55 and 45 days after planting in the main and complementary experiment, respectively. The dominant weed species present in both experiments were *Panicum capillare, Setaria* spp., *Amaranthus retroflexus,* and *Ambrosia artemisiifolia.* Weed biomass in the main experiment ranged from 9 g/m^2 in the 4 species mixture to 200 g/m^2 in the low seeding rate biculture treatment of millet and cowpea. When averaged across all seeding rates, weed biomass in the legume biculture (sunn hemp and cowpea) treatment was 130 g/m^2, whereas weed biomass in the grass biculture (millet and sorghum sudangrass) treatment was only 29 g/m^2. The biculture treatment that produced the greatest crop (sorghum sudangrass and sunn hemp) had a relatively low weed biomass of 41 g/m^2. Preliminary analysis in the complementary experiment shows that weed biomass was greater in monocultures (21 g/m^2) than polycultures (8 g/m^2), with the exception of the millet (6 g/m^2). Overall, the short legume crop, cowpea, was the least weed suppressive, most likely due to less efficient use of sunlight and soil nitrogen. Other the other hand the mixture of sorghum sudangrass, millet, and sunn hemp was one of the most weed suppressive due to greater use of available resources.

RESPONSE OF GRAFTED AND NONGRAFTED TOMATO TO PREEMERGENCE HERBICIDES. S. Chaudhari*, K. Jennings, D. Monks, and F. Louws, North Carolina State University, Raleigh, NC (44)

ABSTRACT

In the United States, interest in the use of grafted tomato as an alternative to the loss of methyl bromide to manage soil borne diseases and pests is increasing. Although several herbicides are registered in tomato production, a lack of information exists on the effect of herbicides on grafted tomato. The use of grafted plants increases cost of production, therefore farmers cannot afford failure of their crop due to herbicide injury. Thus, the objective of this study was to determine the effect of grafting on herbicide tolerance of tomato. Field studies were conducted in 2012 (Mills River, NC) and 2013 (Waynesville, NC). The scion was Amelia tomato and AnchorT, Beaufort, and Maxifort tomato cultivar were used as rootstocks. Herbicide treatments included nontreated, *S*-metolachlor (0.8 and 1.06 kg ai ha^{-1}), fomesafen (0.28 and 0.42 kg ai ha^{-1}), metribuzin (0.28 and 0.55 kg ai ha^{-1}), napropamide (1.12 and 2.24 kg ai ha^{-1}), halosulfuron (0.039 and 0.052 kg ai ha^{-1}), and trifluralin (0.56 and 0.84 kg ai ha^{-1}). Herbicide treatments were applied one day prior to tomato transplanting. The experimental design was a strip-plot with four replications. The whole-plot treatment was rootstock and subplots were herbicide treatments. In 2012, at 1 and 2 wk after treatment (WAT) injury was greater in grafted plants regardless of rootstock type (AnchorT, Beaufort, Maxifort) as compared to non-grafted for halosulfuron, metribuzin, and fomesafen. The maximum injury was observed from halosulfuron on AnchorT (19%). However, by 4 WAT no injury was observed in grafted or non-grafted plants. In 2013, no injury was reported regardless of rootstock and herbicide treatments. Both years no interaction occurred between herbicide and rootstock for height. In 2013, at 8 WAT effect of rootstock on height was significant and Amelia grafted on Maxifort was tallest (103 cm). In 2012, tomato were harvested only once due to bacterial wilt epidemic; herbicide and grafting had no effect on the total fruit yield. In 2013, rootstock had a significant effect on total and marketable yield, and yield was greater with Amelia grafted on Maxifort. These results show herbicides may cause initial injury to grafted plants, however that does not impact the yield. Therefore, these herbicides can be considered effective and safe for weed control in grafted tomato.

NO-HERBICIDE NO-TILL SOYBEAN IN NEW YORK STATE. J. Liebert* and M. Ryan, Cornell University, Ithaca, NY (45)

ABSTRACT

Maximizing biomass of cereal rye (*Secale cereale* L.) has been suggested as being critical for effective weed suppression in cover crop-based organic rotational no-till soybean. However, very little research has been done on other overwintering cereal cover crops or mixtures of cereal cover crops in this system. We conducted an experiment in 2012-2013 in central New York to compare the weed suppression effects of cereal rye and barley (*Hordeum vulgare* L.) in herbicide-free, no-till planted soybean. We hypothesized that a mixture of the two cover crops will result in greater weed suppression than a monoculture of either cover crop. Barley and cereal rye were seeded in a factorial arrangement, allowing us to assess the weed-suppressive ability of each cover crop in monoculture (barley:cereal rye, 100:0 and 0:100), as well as in several bicultures based on seed volume (25:75, 50:50, and 75:25). A split-block design with four replications was used with cover crop and supplemental management as main plots. Supplemental management consisted of a single pass of high-residue cultivation (HRC) with a John Deere 886 approximately 4 weeks after planting. This practice was compared to standard no-till (SNT) management, which did not include high-residue cultivation. Cover crops were terminated with a roller-crimper in mid-June, and soybean (Growmark FS 955BYU13) was no-till planted in 72 cm rows.

Photosynthetically active radiation (PAR) transmittance through the cover crop canopy decreased and total cover crop biomass increased as the proportion of cereal rye seed increased. Average weed biomass ranged from 43 to 127 g m^{-2} across treatments and tended to decrease as the proportion of cereal rye increased. Common ragweed (*Ambrosia artemisiifolia*) and foxtail (*Setaria* species) were the dominant weeds, but they responded differently to supplemental management. Across all barley:cereal rye combinations, common ragweed biomass was 52% lower in treatments that received HRC compared to SNT, whereas foxtail biomass was unaffected or increased with supplemental management. Although we did not find support for our hypothesis, we observed disproportionately lower weed biomass than expected based on cover crop biomass in some mixtures. This suggests that barley might enhance weed suppression through alternative mechanisms, such as increased shading due to different plant statures and leaf architectures via niche partitioning. For example, when comparing the 50:50 to the 0:100 barley:cereal rye mixture, cover crop biomass was 24% lower, but PAR and weed biomass were only 5% and 15% higher, respectively. Although more research is needed, our preliminary results illustrate the weed suppression benefits of combining cultural and mechanical tactics, and suggest that early season shading from cover crops prior to rolling-crimping is an important factor influencing weed suppression.

CHEM-LAWN SPRAY GUN VS. BOOM FOR DITHIOPYR PLUS FLORASULAM PERFORMANCE IN LAWN TURF.

M.C. Cox*, S.D. Askew, and D. Loughner, Virginia Tech, Blacksburg, VA (46)

ABSTRACT

Dithiopyr is a preemergence herbicide commonly used in the transition zone and northern U.S. for crabgrass control. Florasulam is a new broadleaf herbicide from Dow® AgroSciences under the trade name Defendor™ that is currently only sold as a liquid in a co-pack with Dimension® 2EW. Previous research with florasulam documented its ability to suppress bloom production of weeds such as dandelion and white clover. Currently, three-way products containing herbicides such as 2,4-D, MCPP, and dicamba are applied multiple times throughout the spring and summer to control broadleaf weeds. A single application of dithiopyr plus florasulam as a tank-mix could possibly control crabgrass and prolong or even prevent the use of three-way herbicides in cool-season turfgrasses. The objective of this study was to determine the appropriate application timings and tank-mixes of dithiopyr and florasulam for controlling crabgrass, dandelion, and white clover and suppressing their bloom production on a perennial ryegrass lawn with either a Chem-lawn turf spray gun or standard, 4-nozzle research boom.

Smooth crabgrass was not observed at any time during this study, likely due to intense turfgrass competition and increased mowing heights. Herbicide and application main effects or interaction were not significant for the first 8 weeks of the study. At 8 WAIT, dandelion cover, control, and bloom reduction were significantly influenced by herbicide but not by application technique ($p<0.05$). Similar trends were noted for white clover. At 12 WAIT, the herbicide by application technique interaction was significant for dandelion control ($p<0.05$) but only the herbicide main effect was significant for clover control and bloom reduction for both. The interaction was likely due to dithiopyr at the low rate plus florasulam on March 3 followed by May 26 with or without Trimec controlling dandelion better when applied with the spray boom than the Chem-lawn gun. All other herbicides controlled dandelion equivalently regardless of application technique. At 15 and 18 WAIT, only the herbicide main effect was significant for dandelion and white clover control, and bloom reduction. At 15 WAIT, dandelion was controlled most by dithiopyr at 426 g ai/ha plus florasulam on April 28 with or without the addition of Trimec on May 26; however, dandelion control was significantly higher in plots treated with Trimec on May 26 than those without at 18 WAIT. These data suggest that a late spring application of Trimec may prolong dandelion regrowth better than florasulam alone. White clover was controlled by all treatments except dithiopyr applied alone at any rate, 18 WAIT. Based on these data, efficacy of dithiopyr plus florasulam treatments are not hindered by applying with a Chem-lawn spray gun instead of a boom sprayer calibrated for high spray outputs. Herbicide rates and timings were the major influence on dandelion and white clover cover, control, and bloom reduction. Florasulam appears to be effective for dandelion bloom suppression and short-term control; however, a late spring application of Trimec in addition to spring applications of florasulam controls dandelion and inhibits regrowth for a longer duration.

SORGHUM WEED MANAGEMENT AS AFFECTED BY PLANT POPULATION, ROW SPACING AND HERBICIDE PROGRAMS. T.E. Besancon*, W.J. Everman, R. Weisz, R. Heiniger, and R. Riar, North Carolina State University, RALEIGH, NC (47)

ABSTRACT

Weed control remains a major challenge for economically viable sorghum production in North Carolina because sorghum is highly sensitive to weed competition during early growth stages. Moreover, herbicides able to suppress grasses are extremely limited due to sorghum sensitivity. Besides grass weeds, Palmer amaranth (*Amaranthus palmeri*) is one of the broadleaf weeds that may be the most problematic in sorghum production. Previous studies have shown improvements of weed control in sorghum with narrowed row spacing and increased planting density. Separate field studies were conducted in 2012 and 2013 at the Upper Coastal Plain Research Station (Rocky Mount, NC), at the Caswell Research Farm (Kinston, NC), at the Central Crops Research Station (Clayton, NC) and at Clarkton, NC, to determine which row spacings and which plant populations would increase crop competitiveness to allow the reduction of POST herbicide applications. The experiment was conducted in a factorial arrangement of treatments in a randomized complete block design with row spacing (19, 38, and 76 cm), plant population (40,000, 80,000, 120,000, 160,000 plants, and 300,000 plants per acre), and herbicides (non-treated, PRE application of S-metolachlor at 1412 g ai.ha^{-1} + atrazine at 1824 g ai.ha$^{\text{â€‘1}}$, and PRE application of S-metolachlor at 1076 g ai.ha^{-1} + atrazine at 1390 g ai.ha^{-1} followed by POST application of 2,4-D amine at 333 g ai.ha^{-1} or acetochlor at 840 g ai.ha^{-1}) as main factors. Sorghum was rated for the percentage of Palmer amaranth, sicklepod (*Senna obtusifolia*), large crabgrass (*Digitaria sanguinalis*) and crowfootgrass (*Dactyloctenium aegyptium*) control 4 weeks after PRE, and 1, 3 and 7 weeks after POST. Weed density and biomass were evaluated before harvest as well as yield at the harvest.

In 2012, overall, Palmer amaranth density increased with wider row spacings. Its biomass was primarily affected by plant population in 2012 and by both row spacing and plant population in 2013. Herbicide control averaged 98% for both herbicide strategies at Clarkton and Rocky Mount in 2012 and 95% at Rocky Mount in 2013. At Kinston in 2012 and 2013, sicklepod control averaged 40% for PRE herbicide alone and 70% for PRE herbicide followed by POST herbicide. Sicklepod control tended to increase with plant population for the 19 cm row spacings. In 2013 at Clayton, large crabgrass biomass was primarily affected by plant population. Results were similar for crowfootgrass with a decrease of the biomass for both row spacing and plant population. Application of acetochlor as a POST herbicide didn't improve grass control. Sorghum yields were significantly different at Clarkton and Rocky Mount in 2012. The highest yields were associated with the combination of narrow rows and high plant densities.

CURRENT CHALLENGES IN WEED MANAGEMENT IN NORTHEASTERN CRANBERRY PRODUCTION. H. Sandler* and K.M. Ghantous, UMass Cranberry Station, East Wareham, MA (48)

ABSTRACT

In 2013, we established experiments to evaluate control of dodder (*Cuscuta gronovii* and other *Cuscuta* species), moss (*Sphagnum* spp. and *Polytrichum commune*), and poison ivy (*Toxicodendron radicans*). These weeds have had sufficient controls, which have now been lost, or are up and coming pest issues. The industry lost one important herbicide in 2008, pronamide, which provided excellent control of dodder. Since then, quinclorac has exhibited good control in trials in Wisconsin cranberry production. We evaluated quinclorac in MA field studies over the past few years, but have not been able to duplicate Wisconsin's success. Additionally, the use of quinclorac is currently restricted by many handlers due to export (residue) issues.

Moss is viewed as a new emerging pest. We evaluated five products: acetic acid (20% horticultural strength), herbicidal soap (22% ammoniated soap of fatty acids), a mossacidal product (22% potassium salts of fatty acids), hydrogen peroxide (commercial product at 5.3%), and ferrous sulfate monohydrate (17.5%). Applications were made May 15, 2013. The acetic acid and oxidate were applied at full strength. The herbicidal soap was diluted to label rate (204 ml/L). The mossacidal product was applied in a 1:9 dilution. All applications were followed by a light water rinse 15 min after treatment. Ferrous sulfate was applied at the bag label rate (18 g/m^2) and the recommended rate for cranberry (925 g/m^2). All products seem to work well on both types of moss. Acetic acid, herbicidal soap, and oxidate killed the moss but also injured cranberry vines. Application was made after the vines began active growth; phytotoxicity may be reduced with earlier (dormant) applications but the impact on efficacy is unknown. Vines treated with acetic acid recovered within a few weeks. The most promising controls were the low rate of ferrous sulfate and the mossacidal product.

Poison ivy has been a low priority weed for many years but is becoming quite a serious issue on many farms. We evaluated two spot (backpack) applications of a mixture of sethoxydim and mesotrione with a crop oil concentrate (12 ml/L each) at two different timings (first application: May 30 or June 19 plus a follow-up application approximately 2 wk later) on four cranberry farms in Massachusetts. The mixture provided moderate to very good control in the year of application. Plots will be evaluated in 2014.

Little bluestem (*Schizachyrium scoparium*) and broomsedge (*Andropogon virginicus*), also known as poverty grass, are also up and coming serious weed pests for some growers. In 2012, we established 1 by 2m plots on 3 farms in a RCB design with 3 replicates and treated the plots with napropamide (50DF;16.8 kg/ha), sethoxydim (18% a.i.; 15.7 ml/L), or glyphosate (12% solution) with and without subsequent mowing. Plots were also mowed once or twice, handweeded, or left untreated. Biomass was collected in summer of 2013 and sample processing and data analysis are currently underway.

MESOTRIONE: A NOVEL MODE OF ACTION HERBICIDE FOR USE IN TREE NUT, POME FRUIT, STONE FRUIT, AND CITRUS CROPS. . K.D. Burnell*, E. Rawls, L. Glasgow, T. Beckett, R.S. Bounds, J.I. Adkins, and D.L. Hammons, Syngenta, Ontario, NY (49)

ABSTRACT

Mesotrione is a novel herbicide for the tree nut and fruit market that provides systemic preemergence and postemergence weed control. Mesotrione is readily taken up by leaves, shoots or roots, and is translocated in both the xylem and phloem. The mode of action is through competitive inhibition of the HPPD (4-hydroxyphenyl-pyruvate dioxygenase) enzyme which ultimately causes the destruction of chlorophyll, accounting for the typical white "bleaching" symptoms of herbicidal activity. Benefits of the herbicide include: flexibility of pre and post-emergence application timing; low use rates; tank mix flexibility; excellent crop tolerance; excellent efficacy on weed biotypes resistant to other herbicide classes. The use rates range from 105 - 210 g ai/ha applied post-emergence with glyphosate or paraquat, or alone as a pre-emergence application. A maximum of three applications and 420 g ai/ha can be applied per year. Mesotrione is primarily a broadleaf herbicide with significant activity on some grass and sedge species. Key weed species controlled or partially controlled by mesotrione include: pigweeds (*Amaranthus* spp), common lambsquarters (*Chenopodium album*), nightshades (*Solanum* spp), ragweed (*Ambrosia* sp.), morningglory (*Ipomoea* sp.), spanish needles (*Bidens bipinnata*), Asiatic dayflower (*Commelina communis*), hairy fleabane (*Conyza bonariensis*), horseweed (*Conyza canadensis*), large crabgrass (*Digitaria sanguinalis*), broadleaf signalgrass (*Brachiaria platyphylla*), fall panicum (*Panicum dichotomiflorum*), and yellow nutsedge (*Cyperus esculentus*). Field trials have been conducted since 2009 in all major tree crop growing regions. Results indicate that mesotrione alone can provide effective control of many broadleaf weed species, however, overall weed spectrum and control can be improved by the addition of tank mix partners. Mesotrione gives growers another management tool for weed control and when used in combination with glyphosate or paraquat can improve burndown and provide residual control of problematic weeds.

POTENTIAL HERBICIDES FOR ROW-MIDDLE WEED CONTROL IN CANTALOUPE AND WATERMELON IN PLASTICULTURE. D. Lingenfelter*, M.J. VanGessel, B. Scott, and Q. Johnson, Penn State University, University Park, PA (50)

ABSTRACT

Field studies were conducted from 2011 to 2013 at two locations, Georgetown, Delaware (DE) and Rock Springs, Pennsylvania (PA) to examine various potential herbicide programs for row-middle weed control and crop injury in watermelon (var. '7187HQ' and 'Top Gun') and muskmelon (var. 'Aphrodite' and 'Athena') in black plastic. PRE and POST-directed programs were evaluated. Herbicides included: Strategy (0.53 lb ai/A); Curbit (0.56 lb); Sandea (0.03 lb); Chateau (0.096 and 0.128 lb); Sharpen (0.067 and 0.111 lb); Reflex (0.38 lb); Sinbar (0.2 lb); Dual Magnum (1.22 lb); and Gramoxone (0.75 lb). Visual weed control evaluations were taken periodically throughout the growing period. Crop yield data and phytotoxicity ratings were also collected. Small-plot studies were arranged in a randomized complete block design with three replications. Potential herbicide programs study (PA and DE): Chateau and Sharpen have potential to be labeled for weed control in row-middles. These herbicides in combination with Curbit were compared to a PRE standard of Strategy plus Curbit plus Sandea and a POST program of the same mixture plus Gramoxone. In PA, late season ratings revealed that the PRE Strategy tank mixture program provided 88 to 96% control of common purslane, common lambsquarters, redroot pigweed, and hairy galinsoga, while the POST Strategy mixture with Gramoxone provided 97-98% control of the weeds. Both rates of Chateau and Sharpen provided 94-99% and 77-89% control, respectively, of the same species. Across all the treatments, ladysthumb control ranged from 78-86% control. None of the treatments resulted in significant crop injury or yield decreases. In DE, none of the treatments resulted in stunting significantly different than the untreated check for both watermelon and muskmelon. Due to flooding in mid-season the plots were not taken to yield. Late season ratings for common lambsquarters control was excellent (>97%) and morningglory control was >79% for all treatments except Strategy plus Curbit plus Sandea. Fall panicum control was similar for all treatments (averaging 82%), except Sharpen which provided <40% control. Crop injury study in DE: When melon vines were 18-21 inches long herbicide treatments were applied with a hooded sprayer. Residual herbicides (Reflex, Chateau, Dual, Sandea in all three years; and Sinbar and Sharpen in 2011 and 2012) were applied in combination with Gramoxone and a surfactant. In 2011, results were quite variable and no significant differences in treatment were detected. However, watermelon stunting at 8 DAT was 18 and 21 for Chateau and Sharpen, respectively. Although no yield differences were detected, plots treated with Chateau and Dual yield at least 15% less than the untreated check. For muskmelon at 8 DAT, Sharpen, Chateau, and Dual all caused 40, 32, and 20% stunting, respectively. Yield was not significantly different, but Chateau, Dual, Sharpen, and Sinbar all resulted in >15% reduction in yield. In 2012, no treatment was rated significantly different than the untreated check for either watermelon or muskmelon. In 2013 in watermelon, only Reflex had an injury rating (14%) significantly different than the untreated check. No difference in yield was detected. For muskmelon, all treatments resulted in ≥10% stunting at 9 DAT and Reflex, Chateau and Dual treatments were still rated ≥10% stunting 24 DAT. Despite no significant differences for yield, Chateau, Dual and Sharpen treatments resulted in >25% reduction in harvested yield. In this trial, Chateau has the potential to cause stunting to both crops if the spray drift is allowed to injure young plants. Furthermore, muskmelon appears to be more sensitive to Sharpen, Chateau, and Dual than watermelon.

EVALUATION OF PREEMERGENCE HERBICIDE COMBINATIONS TO PREVENT WEED RESISTANCE IN WILD BLUEBERRY FIELDS IN MAINE. D.E. Yarborough* and J.L. Cote, University of Maine, Orono, ME (51)

ABSTRACT

In 2013 we continued to assess three new herbicides for efficacy in weed control and effects on blueberry growth and yield. In April 2013, a trial was set up on nine sites with different soil conditions and weeds to determine the effects of pre-emergence applications of indaziflam, halosulfuron methyl or rimsulfuron on wild blueberry cover/injury and weed cover, alone and combined with hexazinone or terbacil, in a split block experimental design with location serving as replication. The growers' spray regimens with and without halosulfuron methyl was also included, which with the untreated check provided for a total of 14 treatments. An additional location was treated with indaziflam, halosulfuron methyl and rimsulfuron in November 2012 and hexazinone and terbacil were applied at right angles in May 2013 to compare the effects of these three new herbicides when applied in the fall after pruning.

Indaziflam or rimsulfuron, alone or in combination with hexazinone or terbacil, did not reduce wild blueberry cover or result in appreciable phytotoxicity. Halosulfuron methyl, alone or in combination, did show an initial stunting and delay in growth, but the plants recovered fully over the growing season. The growers' spray regimen alone, halosulfuron methyl alone, and the combination had an effect in suppressing broadleaf weeds over the long-term, as did hexazinone, indaziflam+hexazinone, halosulfuron methyl+hexazinone, indaziflam+terbacil and rimsulfuron+terbacil (but was not significant). Grasses were initially controlled by halosulfuron methyl+hexazinone, terbacil, indaziflam+terbacil and rimsulfuron+terbacil. In addition, although in August only rimsulfuron+terbacil continued to control grasses, all of the terbacil combinations continued to maintain grass cover about 5% or below.

Applying indaziflam, rimsulfuron and halosulfuron methyl in the fall improved long-term blueberry cover overall, as well as reduced early halosulfuron methyl phytotoxicity. Fall application improved broadleaf weed control in general compared to spring application when combined with spring hexazinone. Grass control was improved by fall application of indaziflam with spring hexazinone, and to a lesser extent halosulfuron methyl+hexazinone and halosulfuron methyl+terbacil grass control improved as well when compared to halosulfuron methyl applied in spring. Nine new locations are being established in November of 2013 to evaluate more sites for the fall application of indaziflam, halosulfuron methyl or rimsulfuron, alone and combined with spring application of hexazinone or terbacil in 2014 in order to reduce any phytotoxicity from the indaziflam, halosulfuron methyl or rimsulfuron, and improve the efficacy of hexazinone or terbacil by allowing for a later spring application of these materials.

EVALUATING HERBICIDE PROGRAMS FOR ZONE-TILL DRY BEANS. R. Bellinder* and J.B. France, Cornell University, Ithaca, NY (52)

ABSTRACT

Trials were conducted in 2011 and 2012 to evaluate potential herbicide combinations and times of application in conventional [CT] and zone-tillage [ZT] dry beans (*Phaseolus vulgaris* L.). The beans 'Cabernet' were planted into a CT rye cover crop and one that was killed with glyphosate and zone-tilled in 10 inch rows. The herbicide programs were identical in both types of tillage. The application timings were a) preemergence (PRE): metolachlor + fomesafen; pendimethalin + fomesafen; and halosulfuron + fomesafen; b) PRE plus postemergence (POST): metolachlor PRE and bentazon + fomesafen POST; and pendimethalin (PRE) and bentazon + fomesafen POST; c) POST alone: metolachlor + bentazon + fomesafen + clethodim; and pendimethalin + bentazon + fomesafen + clethodim.

The two growing seasons were very different with 2011 being overly wet and 2012 being extremely dry. The wet conditions in the early 2011 season caused slight but transitory crop stunting with the PRE herbicides but did not negatively affect yields. Weed emergence and growth were greater in 2011 than in 2012, and was somewhat better controlled in the CT system than in the ZT system. In 2012 weed control was 90% with all programs in both tillage systems. The PRE program with halosulfuron and fomesafen provided little control of annual grasses in both years. Adding clethodim to the total POST program caused a slight increase in foliar necrosis in the beans however this was temporary and did not affect yields. ZT yields were slightly higher than CT yields in 2011 and the reverse happened in 2012, however, in neither year were there any significant differences. Overall the three types of application programs performed well.

THE IR-4 PROJECT: UPDATE ON WEED CONTROL PROJECTS-FOOD USES. M. Arsenovic*, D. Kunkel, J. Baron, and D. Carpenter, IR-4 Project, Princeton, NJ (53)

ABSTRACT

The IR-4 Project is a publicly funded effort to support the registration of pest control products on specialty crops. The IR-4 Project continues to meet specialty-crop grower's needs for weed control options despite the challenges of a mature market for herbicides and the selectivity of specialty crops to many of the more-recently-introduced herbicides. The Pesticide Registration Improvement Act continues to effect IR-4 submissions and EPA reviews of packages. IR-4 submitted herbicide petitions to the EPA from October 2012 to October, 2013 for: Clomazone on Brassica, head and stem, subgroup 5A Rhubarb and southern pea; Fomesafen on Lima beans; Paraquat on Vegetable, tuberous and corm, subgroup 1C;Sulfentrazone on apple and Pendimethalin on Hops; Onion,bulb subgroup 3-07A, Onion, green subgroup 3-07B, Vegetable, Fruiting group 8-10, Fruit, Citrus group 10-10, Fruit, Pome fruit group 11-10, Fruit, Stone group 12-12, Low growing berry subgroup 13-07G, and Sunflower subgroup 20B. From October 2012 through October 2013, EPA has published Notices of Filing in the Federal Register for: Linuron on cilantro, coriander, dill, parsley and horseradish; Clomazone on Brassica, head and stem, subgroup 5A Rhubarb and southern pea; Fomesafen on Lima beans. EPA established tolerances from October 2012 to October 2013 for: Ethalfluralin on Oilseed group 20;Flumioxazin on globe artichoke, cabbage, Chinese cabbage, olive, pomegranate and prickly pear cactus; Glyphosate on Berry and small frui, group 13-07, carrot, Fruit, citrus, group 10-10: Fruit, pome, group 11-10; Oilseed group 20, except canola; Teff, forage and hay; Vegetable, bulb, group 3-07; Vegetable, fruiting, group 8-10, except okra; sweet poato (increased tolerance), and Vegetable, root and tuber, group 1, except carrot, sweet potato, and sugar beet; Halosulfuron-methyl on artichoke,globe; and Caneberry subgroup 13-07A;Imazosulfuron on Tuberous and corm vegetables, subgroup 1C and melon subgroup 9A; NAA on avocado, Mango, Mamey sapote, rambutan, and Fruit, pome, group 11-10; Prometryn on snap bean, succulent and dill; Quinclorac on Berry, low growing, except strawberry, subgroup 13-07H, and rhubarb and Trifluralin on oilseed group 20 and Teff.

ROUNDUP READY LEARNING XPERIENCE - A NEW TRAINING TOOL. T.D. White*, S. Allen, S. Seifert-Higgins, M. Vigna, and J. Sandbrink, Monsanto Company, St. Louis, MO (54)

ABSTRACT

Many growers are currently seeking new ways to effectively control weeds in soybean and cotton. The Roundup Ready® Xtend Crop System, which is pending regulatory approval, is designed to give farmers another tool to manage glyphosate-resistant weeds and improve their crop yield. Once commercially available, this will be the first time trait and chemistry products will be launched simultaneously as part of the same weed management system. Therefore, proper education and training around effective system use is important. In the summer of 2013, Monsanto successfully held Roundup Ready® Learning Xperience events in major soybean and cotton production areas. These events aimed at educating seed dealers, retailers, and other key stakeholders about the Roundup Ready Xtend Crop System. Sites included various trial types to allow participants to see the performance of the system, including trait, herbicide, and application technology components. In addition, training on the value of basic weed science principles, such as using multiple herbicide modes of action for weed control and how to maximize on-target applications, was integrated into the events. In the future, the Learning Xperience will be expanded to include a broader group of participants.

BICYCLOPYRONE FOR PRE- AND POSTEMERGENCE WEED CONTROL IN CORN.

E. Hitchner*, T.H. Beckett, S.E. Cully, R.D. Lins, J.P. Foresman, and G.D. Vail, Syngenta, Woodstown, NJ (55)

ABSTRACT

Bicyclopyrone is a new selective herbicide for weed control in field corn, seed corn, popcorn and sweet corn. The bicyclopyrone mode of action is inhibition of HPPD (4-hydroxyphenyl-pyruvate dioxygenase) enzyme which ultimately causes the destruction of chlorophyll followed by death in sensitive plants.

Upon registration, SYN-A197 will be the first bicyclopyrone containing product launched with anticipated first commercial application in the 2015 growing season. SYN-A197 is a multiple mode-of-action herbicide premix that provides preemergence and postemergence grass and broadleaf weed control.

Field trials were conducted to evaluate SYN-A197 for weed control and crop tolerance compared to commercial standards. Results show that SYN-A197 very effectively controls many difficult weeds and provides improved residual control and consistency compared to the commercial standards.

ENHANCED WEED MANAGEMENT SOLUTIONS WITH MGI HERBICIDE-TOLERANT SOYBEANS. R. Jain*, D.E. Bruns, T.H. Beckett, B.L. Wilkinson, and B. Erdahl, Syngenta Crop Protection, Vero Beach, FL (56)

ABSTRACT

Field trials were conducted in 2012 and 2013 to evaluate mesotrione-based weed control programs in MGI herbicide-tolerant soybeans stacked with glyphosate tolerance. These multiple mode-of-action herbicide tolerant soybeans enable the use of mesotrione and isoxaflutole pre- and post-emergence in addition to glyphosate and glufosinate-ammonium post-emergence.

Several mesotrione-based herbicide programs provided control of key weed species, including glyphosate resistant populations. The most successful and consistent weed control was achieved with two-pass programs that included pre-emergence residual herbicides and multiple, overlapping modes of action. These programs were designed to align with HRAC principles of weed resistance management. The use of these chemically diverse and novel programs will offer effective, safe and sustainable weed management options for soybean growers.

BICYCLOPYRONE, A NEW HERBICIDE FOR IMPROVED WEED CONTROL IN CORN. G.D. Vail*, S.E. Cully, R.D. Lins, and J.P. Foresman, Syngenta, Greensboro, NC (57)

ABSTRACT

Bicyclopyrone is a new selective herbicide for weed control in field corn, seed corn, popcorn and sweet corn. The bicyclopyrone mode of action is inhibition of HPPD (4-hydroxyphenyl-pyruvate dioxygenase) enzyme which ultimately causes the destruction of chlorophyll followed by death in sensitive plants. Bicyclopyrone is safe when applied either pre or postemergence to corn and provides pre and postemergence control of grass and broadleaf weeds.

Upon registration, SYN-A197 will be the first bicyclopyrone containing product launched with anticipated first commercial application in the 2015 growing season.

COVER CROP ADOPTION IMPACTING HERBICIDE SELECTION WITH POTENTIAL CARRYOVER CONCERNS. W. Curran* and D. Lingenfelter, Penn State, University Park, PA (58)

ABSTRACT

Over the last several years, we are increasingly asked about herbicide persistence and cover crop establishment. Our group has attempted some research focused on herbicide carryover particularly related to fall cover crop establishment after corn silage. We conducted herbicide-cover crop trials in 2010 and 2011 at our Rock Springs Agronomy farm in Central Pennsylvania. In both years, corn was harvested for silage in September and a number of annual species including winter cereals, brassicas, and legumes were no-till seeded. A few cover crops were inter-seeded into corn in mid June using the Penn State Interseeder. To make a long story short, the result showed that metolachlor PRE and nicosulfuron POST can reduce annual ryegrass stand particularly when inter-seeded, but many herbicides did not affect the cover crops. The results from these field experiments suggested that many herbicides believed to be problematic were not. However, because of some problems we encountered in conducting this research, the lack of negative results, and the small inference space (only two years at a single location), we pursued other opportunities to demonstrate these relationships. In 2012 and 2013, we conducted nonreplicated demonstrations of herbicide selectivity for a number of different herbicides or combinations. In these demonstrations the fields were tilled and cover crops including legumes, grasses, and brassicas were seeded again using a drill. Herbicides were applied PRE perpendicular to the direction of cover crop seeding and the trials were irrigated to ensure herbicide activation. We utilized a log sprayer for these demonstrations applying the herbicides at rates ranging from 1/16X to 1X or the normal rate. An individual plot and accompanying herbicide rate was about 3 by 6 m in size and included 6 to 8 different crop species. We evaluated the plots for % stand reduction about 4 weeks after establishment and conducted field days at about the same time. At the field days, we utilized the half-life values from the literature along with the herbicide stand reduction ratings to discuss the potential for carryover injury based on herbicide and application rate. Each reduction in herbicide rate represented a half-life of the herbicide and we were able to apply rates theoretically up to 4 half-lives. This technique was very useful especially for longer-lived herbicides but also showed positive results for herbicides with less soil activity. Short-lived herbicides with high soil activity (e.g. rimsulfuron) were not accurately represented in this demonstration since they impacted plant growth at very low rates, yet are short-lived. In future research, we would like to bridge herbicide persistence research with practical demonstration activities that better show how soil persistence, activity, and susceptibility are all involved in predicting rotational crop safety.

PALMER AMARANTH MANAGEMENT IN HPPD, GLUFOSINATE, AND GLYPHOSATE-TOLERANT SOYBEAN. W.J. Everman*, M. Rosemond, and J. Allen, North Carolina State University, Raleigh, NC (59)

ABSTRACT

Glyphosate-resistant Palmer amaranth is the greatest weed management issue for Southeastern soybean producers. Farmers in North Carolina often design their production plans with Palmer amaranth management in mind. Bayer CropScience is developing soybean tolerant to HPPD-inhibiting herbicides in order to provide alternative control options. Research was conducted in North Carolina in 2012 and 2013 to investigate crop tolerance and efficacy in HPPD tolerant soybeans. Two studies were conducted in Clayton, NC to investigate tolerance of HPPD tolerant soybeans to PRE and POST applied HPPD inhibitors and to determine efficacy of weed management programs based on these herbicides. In the first trial, fifteen treatments consisting of PRE fb POST herbicide applications were compared, and in the second trial seventeen PRE fb POST or POST only treatments were evaluated. In both studies excellent crop tolerance was observed. Palmer amaranth control was greater than 95% for all sequential treatments, with residual HPPD-inhibiting herbicides providing an excellent foundation for subsequent POST applications. Results affirm the need for residual herbicides in a comprehensive weed management program for Palmer amaranth.

DICAMBA FORMULATION ADVANCEMENTS. D.J. Mayonado*, J. Sandbrink, A. Macinnes, J.W. Hemminghaus, J.N. Travers, S. Seifert-Higgins, and S.E. Curvey, Monsanto, Salisbury, MD (60)

ABSTRACT

Monsanto Company is developing formulations containing dicamba for use in the Roundup Ready® Xtend™ Crop System. A premix formulation containing diglycolamine (DGA) dicamba and monoethanolamine (MEA) glyphosate delivering a 2 to 1 ratio of glyphosate to dicamba has been developed. A dicamba standalone formulation based on the DGA dicamba salt has also been developed. Both formulations contain proprietary VaporGrip™ technology that reduces the potential of dicamba volatility compared to dicamba formulations that do not. These formulations show commercially acceptable physical/ chemical properties consistent with Roundup® agricultural herbicide formulations and are pending regulatory approval.

Although volatility is a small contributor to potential off-target movement, this often remains a concern from growers and applicators as a legacy from the dimethylamine (DMA) salt launched in the 1960s. The DGA salt of dicamba consistently shows lower volatility and this can be reduced further by using VaporGrip™ technology. Spray drift and tank contamination are the main contributors to potential off-target movement. These can be decreased significantly through appropriate application practices and proper tank clean out. Application requirements for both have been developed as part of the Roundup Ready® Xtend™ Crop System to increase on-target applications.

Commercialization is dependent on multiple factors, including successful conclusion of the regulatory process. The information presented herein is provided for educational purposes only, and is not and shall not be construed as an offer to sell, or a recommendation to use, any unregistered pesticide for any purpose whatsoever. It is a violation of federal law to promote or offer to sell an unregistered pesticide.

EARLY SPRING APPLICATIONS OF INDAZIFLAM ON CONTAINER GROWN ORNAMENTALS. S. Barolli*, Imperial Nurseries, Granby, CT (61)

ABSTRACT

In our trial at Imperial Nurseries last year we learned that sprays of indaziflam have the potential to be safely used on some container grown plants. With the objective for new sprayable herbicides that can be used over the top of those plants we decided to look on early spring applications of indaziflam as e new tool for better weed management.

Three plants of seven species were included. Plants were; rhododendron (Catawbiense 'Grandiflorum'), Gold Thread Cypress (Chamaecyparis pisifera 'Filifera Aurea)', Blue Star Juniper (Juniperus squamata 'Blue Star)', Spirea ('Magic Carpet)', Rose ('Double Knock Out'), Potentilla (Fruticosa 'Abbottswood') and Lilac (Syringa vulgaris 'Fiala Remembrance'). All plants were potted in summer of 2012 and were applied with OH-2 herbicide in October 2012.

Four different herbicide treatments and a control were replicated three times in randomized complete blocks. The treatments were sprays of 1) isoxaben (Gallery 75 DF) at 1 lb/A plus prodiamine (Barricade 4FL) at 21 oz/A), 2) indaziflam (Marengo) at 9 oz/A, 3) indaziflam (Marengo) at 18 oz/A and 4) dimethenamid-P (Tower) at 26 oz/A. Sprays were done with a Solo backpack equipped with an 8004-VS nozzle operating at 30 PSI and 3 ft/sec, applying 40 gal/A. Applications were done on March 7th 2013. Plants did have very small leaves, except for Roses which were in full vegetation. The house was covered with white plastic and doors were open the day of applications. The temperature was 36°F. Plants were irrigated for 5 minutes prior to applications and 15 minutes after all sprays were done. Plants were held in the hoop house for more than two months. Three ratings of injuries were done, respectively two weeks, one month and two months after applications. The scale was (0-10); zero-no injury and ten- dead. Efficacy for weed control was done 2 months after the applications.

None of the treatments injured Blue Star and Gold Thread. Except Marengo at 18 oz/A, no treatment injured rhododendron and those injured had very little injury after 1 month and grew out if it. Rose, lilac and potentilla were injured with both rates of Marengo and Tower. Spirea and potentilla did have growth reduction by sprays of Marengo. Rose was injured by Gallery + Barricade. All treatments had good control on the weeds and no significant differences between them.

The results of this and previous experiments show that indaziflam sprays can be safely used in some container grown ornamentals in early spring. Based on those results research is needed to see the potential of this herbicide over the top of other container grown plants.

RICE HULLS FOR WEED CONTROL IN CONTAINER CROPS. J. Altland*, USDA-ARS, Wooster, OH (62)

ABSTRACT

Preemergence herbicides are the primary tool used for weed management in container crops. Weed control alternatives are needed for crops that are either sensitive to preemergence herbicides, or in sites where preemergence herbicides are not labeled. Rice hulls are commercially available to the horticulture industry, inexpensive, and potentially useful for weed control in container crops. The objective of this research was to determine the efficacy of rice hulls for controlling liverworts (*Marchantia polymorpha* L.) and bittercress (*Cardamine flexuosa*) in a greenhouse environment with or without the application of fertilization through an irrigation system.

On March 20, 2013, 15-cm diameter pots were filled with Sunshine LB2 potting mix (Sun Gro Horticulture Canada Ltd., Seba Beach, AB, Can.). Containers were filled to within 2.5 cm of the container top. On March 21, 2013, containers were randomly assigned to receive rice hulls at a depth of 0, 0.63, 1.25, or 2.5 cm by weighing 0, 11, 22, or 44 g of rice hulls and spreading them evenly over the surface. Half of the containers in each rice hull treatment were placed on a bench receiving overhead irrigation of city tap water, and the other half on a bench receiving overhead irrigation of city tap water with 100 mg/L N from a commercial fertilizer (20N-4.3P-16.6K, JR Peters, Inc., Allentown, PA). Irrigation was run twice daily for 7 min, so that each plot received approximately 1 cm water per day. Approximately 20 liverwort gemmae were applied to the surface of each container weekly. Ten bittercress seed were applied to the surface of each container at the initiation of the experiment (after rice hulls were applied) and again at 4 weeks after potting. Data collected included a visual estimation of the percent of container surface covered by liverwort thalli at 2, 4, 6, and 8 weeks after potting (WAP). At 4 and 8 WAP, bittercress were harvested and shoot fresh weights (SFW) per pot were measured.

Liverwort coverage of the container surface was greater in containers receiving fertilization compared to those that did not (Table 1). Within a fertilizer treatment, liverwort was greatest in control containers that received no rice hulls. Containers receiving 0.63 cm of rice hulls were similar to those receiving 1.25 or 2.5 cm on most dates. Despite this, some liverwort grew in all containers receiving 0.63 cm of rice hulls, while no liverwort growth was observed in those receiving either 1.25 or 2.5 cm. A similar trend with respect to fertilization and rice hull depth was observed with bittercress SFW. Data in Table 1 show no bittercress SFW in containers that received irrigation only (no fertilizer). However, bittercress had germinated in those containers but never developed beyond the cotyledon stage and thus had insufficient mass for measurement.

It was hypothesized that rice hulls would not provide effective weed control in containers fertilized through the irrigation system due to the constant source of nutrition available to germinating seed and gemmae. However, it was observed that rice hulls dried very quickly after irrigation events, making establishment of liverwort or bittercress difficult. Application of 0.63 cm rice hulls was not sufficient to completely cover the substrate surface. This depth of rice hull allowed for some gaps where the substrate surface was exposed, and it was these sites that allowed for weed establishment. This research demonstrates that rice hulls are effective in controlling liverwort and bittercress in production systems utilizing constant fertilization injected through the irrigation system.

Table 1. Percent of the container surface covered by liverwort and bittercress shoot fresh weight in containers that were either fertilized or not, and receiving parboiled rice hull mulch from 0 to 2.5 cm in depth. The fertilizer applied to selected treatments was a complete commercial fertilizer (20-10-20) injected in the irrigation system. WAP=weeks after potting.

		liverwort coverage				bittercress fresh weight	
Fertilizer	rice hull depth	2 WAP	4 WAP	6 WAP	8 WAP	4 WAP	8 WAP
	cm	%	%	%	%	g	g
No	0.00	2	4	5	2	0.0	0.0
No	0.63	0	1	3	5	0.0	0.0
No	1.25	0	1	1	0	0.0	0.0
No	2.50	0	0	0	0	0.0	0.0
Yes	0.00	12	27	31	29	1.2	5.8
Yes	0.63	2	4	10	8	0.1	0.8
Yes	1.25	0	0	0	0	0.0	0.0
Yes	2.50	0	0	0	0	0.0	0.0
LSD (p=0.05)		3	9	9	9	0.4	1.4

LIVERWORT CONTROL IN NURSERY CROPS. J. Derr*, Virginia Tech, Virginia Beach, VA (63)

ABSTRACT

An important weed species in plant propagation as well as in production of perennials is liverwort (*Marchantia polymorpha* L.). Liverwort is a primitive plant that forms leaf-like mats on the soil surface and grows best in cool, moist conditions. Propagation in enclosed structures, which limits herbicide use, makes control a particular challenge. The objective of this research was to identify preemergence and postemergence chemicals that effectively control liverwort.

The preemergence study evaluated common granular formulations used in the nursery trade. The most effective herbicide was dimethenamid + pendimethalin, which provided 84 and 89% control 70 DAT at 1.75 and 3.5 lb ai/A, respectively. Flumioxazin at 0.38 lb/A and oxyfluorfen + oryzalin at 3.0 lb/A both provided approximately 75% control 70 DAT. Oxadiazon at 4.0 lb/A gave 50% control.

In the postemergence trial, oregano oil at 2% v/v and pelargonic acid at 5 and 10% v/v caused significant and rapid injury (80% or higher) to liverwort within 30 min of application. Effects from most of the other treatments were apparent 1 day after treatment (DAT). At 1 DAT, oregano oil, pelargonic acid, and acetic acid (10 and 20%) all provided 75% or greater control of liverwort. By 21 DAT, however, only sprayed applications of flumioxazin at 0.38 lb/A and dimethenamid at 1.5 lb/A gave greater than 85% control as regrowth occurred in the other treatments, with pelargonic acid, acetic acid and 2% oregano oil providing 60% or greater control. A second application was made at 3 weeks after the initial treatment. At 15 minutes after the second application, oregano oil, ammonium nonanoate at 5%, pelargonic acid, flumioxazin, dimethenamid, and acetic acid all gave 75% or greater liverwort control. These treatments all provided 80% or greater control at 8 days after the second application. The 2.5% rate of ammonium nonanoate and both granular and sprayed applications of sodium carbonate peroxyhydrate did not provide acceptable control of liverwort. Thorough coverage of liverwort appears to be critical for all of these treatments since the action seems to be contact for each one.

Another trial compared rates of sprayed dimethenamid and iron HEDTA, as well as granular treatments of dimethenamid + pendimethalin. Dimethenamid at 1.5 lb/A gave better POST liverwort control than at 1.0 lb/A. At 20 DAT, dimethenamid at 1.5 lb/A gave 95% liverwort control while the higher rate of iron HEDTA (10 gal/1000 sq. ft.) gave 66% control. The granular formulation of dimethenamid + pendimethalin suppressed but did not provide acceptable control. Since the dimethenamid formulation contains an oil solvent, an experiment was conducted to determine if it was the oil solvent that was providing the liverwort control. The formulation blank injured but did not provide significant control. So the control was due either to the herbicide itself or a synergistic action of the herbicide with the solvent.

An experiment evaluated the impact of irrigation immediately after treatment on liverwort control. In general, irrigating after application reduced crop injury but also reduced liverwort control.

Several chemicals, including oregano oil, acetic acid, pelargonic acid, flumioxazin, and dimethenamid, provide acceptable liverwort control. Additional research is needed to maximize control while minimizing crop injury, especially for the nonselective chemicals.

COMPARATIVE LONGEVITY OF PREEMERGENCE WEED CONTROL IN CONTAINERS WITH INDAZIFLAM. J. Neal* and C. Harlow, North Carolina State University, Raleigh, NC (64)

ABSTRACT

The objective of this experiment was to compare the longevity of preemergence weed control in containers with Marengo G herbicide, relative to industry standard herbicides. The experiment was conducted at the Horticultural Crops Research Station in Castle Hayne, NC. On April 30, 2013, 4-Liter plastic pots were filled with a pine bark + sand substrate (10:1 by volume) amended with Osmocote Pro 17-5-11 slow release fertilizer @ 6#/yd^3. Pots were hand watered to settle the substrate then received 0.6 inches of overhead irrigation for about two weeks to ensure substrate had settled and fine particulates had stabilized. Treatments were applied on May 14, 2013 then pots received 0.6 inches of overhead irrigation. Treatments included Marengo 0.0224% G (indaziflam) at 100, 150 and 200 lb /A, Broadstar 0.25G (flumioxazin) at 150 lb/A, Everris OH2 (formerly Scotts OH2) (oxyfluorfen + pendimethalin) at 100 lb/A and Snapshot 2.25 G (isoxaben + trifluralin) at 200 lb/A, as well as a non-treated control. Treatments were arranged in a randomized complete block design with main plots split into sub-plots of 7 seeding regimes with separate pots seeded 0, 2, 4, 6, 8, 10, and 12 weeks after treatment. Weeds were surface seeded to achieve about 40 viable seeds per pot with 3 pots of each species per sub-plot. Weed species included *Digitaria sanguinalis, Cardamine flexuosa, Chamaesyce maculata,* and *Eclipta prostrata*. Percent weed control compared to the non-treated pots was visually evaluated 2 and 4 weeks after each seeding event. Bittercress and spurge were controlled for a longer period of time than were eclipta or crabgrass. Marengo G provided greater than 80% bittercress control for over 10 weeks. The next best treatment on bittercress was Broadstar which provided 6 weeks of greater than 80% control. It should be noted that in previous experiments, Broadstar and OH2 provided about 10 weeks of bittercress control. Marengo at 100 lb/A, Broadstar at 150 lb/A, and OH2 at 100 lb/A provided equivalent control of spurge. Longevity of control improved with increasing dose of Marengo with greater than 80% control 10 weeks after treatment with 200 lb/A Marengo G. Marengo G provided better eclipta control compared to Broadstar, OH2 and Snapshot. Longevity of eclipta control increased with increasing dose from 100 to 150 lb/A. No improvement was observed when dose was increased to 200 lb/A. These data demonstrated that Marengo G applied at $\geq$ 150 lb/A provided longer residual control of spurge, bittercress and eclipta compared to currently labeled industry standard herbicides. Increasing Marengo G dose from 100 to 150 lb/A significantly improved overall weed control and longevity of control. While increasing dose from 150 to 200 lb/A did improve longevity of spurge control, little or no improvement was observed in control of the other three species tested.

TITRATED RATES OF DIMETHENAMID-P FORMULATIONS CONTROL COOL- AND WARM-SEASON WEED SPECIES. K.E. Kalmowitz* and J. Neal, BASF Corp., Raleigh, NC (65)

ABSTRACT

Dimethenamid-p is a preemergence herbicide used for weed control in nursery production, landscapes and in turfgrass use sites. The active is formulated as a combination granular product that includes 1% of pendimethalin and 0.75% dimethenamid-p and as a solo liquid product formulated as a 720 g/l emusifiable concentrate. The active ingredient has a range of efficacy on grass, broadleaf and sedge species. Rotation of herbicides with different mode of action is essential in nursery production and landscapes; however, plant safety sometimes plays the major role in selection of the best products to use at any given time of the year. Both the liquid and granular formulations containing dimethenamid-p were screened at titrated rates for control of a range of weed species that represent the diversity of species that occur in the cool and warm seasons of the months. Control was evaluated as compared to an industry standard in trials conducted over a two year period in more than one state. Differential response to formulations and rates were observed to the following species: Hyssop spurge (*Chamaesyce hyssopifolia*), Hairy bittercress (*Cardamine hirsuta), Large crabgrass (Digitaria sanguinalis),* Sowthistle *(Sonchus* spp.*),* Longstalked phyllanthus (*Phyllanthus tenellus*), Eclipta (*Eclipta alba*), Asiatic hawksbeard (*Youngia japonica*), Dogfennel (*Eupatorium capillifolium*), sedge spp. (*Cyperus* spp.) and Galinsoga spp. (*Galinsoga* spp.).

LIVERWORT MANAGEMENT WITH POSTEMERGENCE HERBICIDES IN CONTAINER GROWN ORNAMENTALS. A. Senesac* and I. Tsontakis-Bradley, Cornell Cooperative Extension, Riverhead, NY (66)

ABSTRACT

In 2013, a trial was conducted under USDA's IR-4 Program sponsorship to determine the crop tolerance and efficacy of organic and biopesticides used for liverwort (*Marchantia polymorpha* L.) management. Treatments were applied in a covered unheated hoop house during the winter and again during active growth of an assortment of container-grown perennials that are subject to infestations of liverwort. Containers of liverwort were also evaluated for efficacy. In one half of the plots, plants were irrigated shortly after treatment application to determine the potential loss of efficacy and the safening effect of "washing off" the products. In a trial conducted in 2012, the need for additional information regarding the efficacy of a single treatment timing was evident. In 2013, one set of liverwort containers was treated at both timings, while another set was treated at the dormant timing only. If a single dormant timing yielded satisfactory control, the need for the spring timing, which has potential to be more injurious to the crop, would be eliminated. Treatments were applied on March 1, 2013 when ornamentals were dormant and on April 26, 2013 when they were actively growing. They included d-limonene, oregano oil extract, and ammonium nonanoate at two rates each, with and without post-treatment irrigation. Test plants included Sedum spurium 'John Creech', Ophiopogon japonicus 'Dwarf Black', and Alchemilla erythropoda 'Alma'. Liverwort had grown to approximately 75% ground cover at the first treatment and 100% cover in untreated containers at the second timing. Phytotoxicity and efficacy data were collected at 1, 2, and 4 weeks after each treatment and 8 weeks after the first treatment, immediately preceding the second application. Crop injury was evident for most treatments after the first application timing with complete or near complete recovery by 8WAT. By the end of trial, most plants had recovered somewhat or completely from injury. A notable exception was Ophiopogon, for which oregano oil extract was very safe and d-limonene from which recovery was not observed. The higher treatment rates and treatments that did not receive post-treatment irrigation were more injurious to the crop. The same effect was not evident for liverwort control. Wash-off did not always reduce efficacy significantly. Likewise, treating liverwort at the second timing did not always improve control, especially at the low rates. For example, d-limonene was very effective- providing 70% control at the low rate with post-treatment irrigation and 90% control at the two high rate treatments. Treatments of d-limonene at the second timing did not improve control significantly.

UPDATE ON 2013 WEED SCIENCE RESEARCH IN THE IR-4 ORNAMENTAL HORTICULTURE PROGRAM. C. Palmer*, J. Baron, E. Vea, and E. Lurvey, IR-4 Project, Princeton, NJ (67)

ABSTRACT

In 2013, the IR-4 Ornamental Horticulture Research Program sponsored research on two weed science projects: 1) crop safety of over-the-top herbicide applications, and 2) crop safety and efficacy of over-the-top herbicides targeting liverwort (*Marchantia polymorpha* L.). The goal of the over-the-top herbicide application evaluations was to screen herbicides [Biathlon (oxyfluorfen + prodiamine), F6875 4SC (sulfentrazone + prodiamine), Freehand G (dimethenamid-p + pendimethalin), Gallery 75DF (isoxaben), Marengo 0.0224%G (indaziflam), and Tower EC (dimethenamid-p)] for safety on woody and herbaceous perennials grown primarily in container nurseries. Applications were made at dormancy and approximately 6 weeks later for all products. Biathlon was applied to 23 crops; F6875 4SC was applied to 32 crops; Freehand was tested on 21 crops; Gallery was applied to 7 crops; Marengo was tested on 9 crops and Tower was applied on 39 crops. The goal of the liverwort herbicide product evaluations was to determine whether immediate irrigation after application reduces crop injury while maintaining efficacy with herbicides effective for managing liverwort. Seven products were evaluated: Avenger Ag (d-limonene), Bryophyter (oregano oil), Racer Herbicide (ammonium nonanoate), Scythe (pelargonic acid), Sureguard 51WDG (flumioxazin), Tower (dimethenamid-p), and Weed Pharm (acetic acid). Applications were made to pots containing liners with or without liverwort followed by irrigation or no irrigation. Data on efficacy and crop safety were collected. With results received thus far, non-selective herbicides applied over the top generally provided good to excellent efficacy, but commercially unacceptable injury was observed even with immediate irrigation on several species. The results from this research will aid in the development of product labels and will help growers and landscape care professionals make more informed product choices.

BENTGRASS RESPONSE TO TOPRAMEZONE. K.A. Venner*, S.D. Askew, M.C. Cox, and K. Miller, Virginia Tech, Blacksburg, VA (68)

ABSTRACT

Topramezone (Pylex™) is an herbicide from BASF newly marketed in turfgrass for control of weedy annual grasses and broadleaves. Topramezone is currently not registered in creeping bentgrass but recent studies indicate that lower rates of topramezone may control selected weeds and suppress bermudagrass while causing only slight or transient injury to bentgrass turf. The objective of these studies was to determine the tolerance of creeping bentgrass to topramezone rates. A field trial was arranged in a randomized complete block design with 3 replications on a 9 year old, 'L-93' fairway maintained at 1.9 cm. Plot size was 1.8 m^2, and applications were made with a CO_2 powered backpack sprayer equipped with 11004 TTI nozzles and calibrated to deliver 280 L ha^{-1} at 262 kPa. Treatments were arranged in a factorial design with four levels of topramezone rate (6.2, 12.3, 18.4 and 24.6 g ae ha^{-1}) and three levels of seasonal initial application [June 3 (spring, SP), July 1 (early summer, ES), and August 15, 2013 (late summer, LS). All herbicide treatments contained methylated seed oil at 0.5% v/v and were applied three times at 3 week intervals. To investigate the influence of sequential treatments, each sequential treatment was considered a main plot in a split plot treatment design and the factorial arrangement of rate and season were considered subplots. Data were analyzed in SAS version 9.1 using the general linear models procedure.

The interaction of sequential treatment by season by rate was significant for turfgrass injury at 1 and 3 weeks after treatment (WAT) ($P<0.05$). In SP, turfgrass injury increased 7.5 to 10% with each 2.5 g ae ha^{-1} increase in topramezone rate. All rates above 0.25 fl oz/A injured creeping bentgrass unacceptably. In ES and LS, the first and second applications injured creeping bentgrass less than in SP. Only SP treatments injured creeping bentgrass at 3 WAT. In ES and LS, creeping bentgrass recovered from initial injury before 3 WAT, regardless of rate. The data suggest that repeated treatments in spring can lead to cumulative injury of creeping bentgrass while creeping bentgrass treated in summer is more likely to recover after each treatment. Spring rains in 2013 could have increased bentgrass response to topramezone. Other HPPD-inhibiting herbicides, like mesotrione, have been known to cause more severe symptoms on desirable species when soil moisture and humidity are elevated. Decreased injury in summer could be the result of reduced turfgrass growth during summer stress and decreased rainfall over time. These data suggest topramezone can be applied safely to creeping bentgrass turf, but rates should not exceed 6.2 g ae ha^{-1} during prolonged rainy periods to avoid turfgrass thinning as well as undesirable turfgrass bleaching following application during normal conditions. Rates lower than 6.2 g ae ha^{-1} could be effectively utilized on fairway height creeping bentgrass with acceptable levels of injury and rapid recovery.

THE MOBILE WEED MANUAL: A NEW RESOURCE FOR MANAGERS OF TURFGRASS AND ORNAMENTALS. M. Elmore*, J.T. Brosnan, G.R. Armel, G.K. Breeden, and J.J. Vargas, University of Tennessee, Knoxville, TN (69)

ABSTRACT

Mobile Weed Manual (MWM) is a new Extension resource developed to assist green industry professionals in selecting herbicides for use in turf and ornamentals. MWM is a web-based platform optimized for use on mobile devices such as smartphones and tablet computers. The site provides users with weed control efficacy information for more than 80 different herbicides, tolerance information for over 2300 turf and ornamental species, and direct links to label and material safety data sheet information for herbicides used for turf and ornamental weed management. MWM provides users with several search options for selecting herbicides for PRE or POST weed control. Users can indicate the desirable turf or ornamental species in which weeds are present and MWM will provide a list of herbicides labeled for use on that desirable species. Moreover, users can further refine searches by indicating both a desirable species and target weed; MWM will provide a list of herbicides labeled for control of the target weed in the desirable species indicated. Since debuting online in May 2013, MWM has allowed Extension specialists to reach over 10,000 individuals from 72 different countries, all 50 United States, and 142 different municipalities throughout Tennessee.

ETHEPHON ABSORPTION AND DISTRIBUTION ATTRIBUTED TO ANNUAL BLUEGRASS SEEDHEAD SUPPRESSION. S. Sidhu* and P. McCullough, University of Georgia, Griffin, GA (70)

ABSTRACT

Ethephon suppresses annual bluegrass (*Poa annua* L.) inflorescence, but physiological effects attributed to efficacy have received limited investigation. Experiments were conducted to evaluate application placement, absorption, distribution, and metabolism of ethephon associated with inflorescence suppression in annual bluegrass. Ethephon applications to the antepenultimate leaf reduced inflorescence length 38 to 63% from the nontreated, while applications to the flag leaf reduced length 67 to 93% from 7 to 28 d after treatment. Annual bluegrass treated with ^{14}C-ethephon at the flag leaf distributed ≈7x more radioactivity to the inflorescence than applications to the antepenultimate leaf, and ≈2x more ^{14}C into nontreated shoots after 72 h. Radioactivity distribution to roots was minimal (≤2%) for both application placements. Annual bluegrass reached 50% foliar absorption of ^{14}C-ethephon at 13 and 17 h after treatment to the antepenultimate and flag leaf, respectively. Annual bluegrass metabolized ^{14}C-ethephon after 1 d to a nonpolar metabolite in the flag leaf, but a different metabolite accounted for ≥88% of ^{14}C recovered after 7 d. Results suggest ethephon absorption and transport from the flag leaf contributes to annual bluegrass inflorescence suppression more than lower leaves, while transport from the flag leaf is likely in the form of a metabolite.

EFFICACY OF DITHIOPYR FOR POSTEMERGENCE SMOOTH CRABGRASS CONTROL IN WARM AND COOL-SEASON GRASSES. H. Jordan* and P. McCullough, University of Georgia, Griffin, GA (71)

ABSTRACT

Dithiopyr is a pyridine herbicide that provides pre- and early postemergence control of smooth crabgrass in turfgrass. The objective of this research was to evaluate formulation and application timing on efficacy of dithiopyr for smooth crabgrass control in bermudagrass and tall fescue. Experiment 1 was conducted on a common bermudagrass field, and Experiment 2 was conducted on a 'Talladega' tall fescue field. Bermudagrass and tall fescue were mowed weekly during active growth at 5 and 7.5-cm heights with a rotary mower, respectively, and irrigated as needed to prevent wilting. For the tall fescue experiment, treatments included dithiopyr 2 EW at 0.56 kg a.i. ha^{-1}, dithiopyr 0.27 GR at 0.56 kg a.i. ha^{-1}, and quinclorac at 0.84 kg a.i. ha^{-1}. For the bermudagrass study, treatments included dithiopyr 2 EW at 0.56 kg ha^{-1}, dithiopyr 0.27 GR at 0.56 kg ha^{-1}, and sethoxydim at 0.28 kg a.i. ha^{-1}. Treatments were applied at three growth stages of smooth crabgrass including multi-leaf, 1 to 2 tiller, and 3 to 5 tiller. An untreated check was included in both experiments. In the tall fescue experiment, all treatments provided ≥88% control of multi-leaf and 1 to 2 tiller smooth crabgrass in late June. By late August, dithiopyr 2EW provided good control (>80%) of smooth crabgrass at the multi-leaf timing, but control from tillered timings was poor (<70%). Dithiopyr 0.27 GR provided fair control (70 to 79%) of smooth crabgrass by late August at the multi-leaf and one-tiller timing, but treatments at multi-tiller timing provided poor control. By late August, quinclorac treatments in tall fescue at the multi-leaf treatment provided poor control, while treatments at the 1 to 2 tiller and 3 to 5 tiller timings provided fair control. In the bermudagrass experiment, both dithiopyr formulations controlled smooth crabgrass 78% in June, but control was poor by late August. Dithiopyr treatments provided poor control of tillered smooth crabgrass in June and August. Sethoxydim excellent control of multi-leaf and 1 to 2 tiller smooth crabgrass in June but control declined to 71% by August. Sethoxydim provided fair control of 3 to 5 tiller smooth crabgrass in June, but control was poor by August. Dithiopyr and quinclorac treatments did not significantly injure tall fescue. Dithiopyr did not significantly injure bermudagrass, but sethoxydim caused unacceptable injury (>20%) on several dates.

EVALUATION OF F6875 FOR SMOOTH CRABGRASS CONTROL IN BERMUDAGRASS AND TALL FESCUE. C. Johnston*, P. McCullough, and B. Walls, University of Georgia, Griffin, GA (72)

ABSTRACT

F6875 alone and on fertilizer carriers are under evaluation by FMC for preemergence control of summer annual weeds in turfgrass. Field experiments were conducted in Griffin, GA on bermudagrass and tall fescue to evaluate efficacy of F6875 formulations for preemergence control of smooth crabgrass and turfgrass tolerance. Bermudagrass and tall fescue were irrigated, and mowed at 1.6-cm 2 d wk^{-1} and 7.5-cm weekly, respectively. Treatments were made on March 26, 2013 to a 'Tifway' bermudagrass fairway, and on March 27, 2013 to a 'Titan' tall fescue lawn. Smooth crabgrass ground cover increased in the nontreated plots from March to August, and measured 20% and 71% after 5 months in bermudagrass and tall fescue, respectively. None of the F6875 formulations injured bermudagrass or tall fescue. At 5 months after treatments in bermudagrass, F6875-9 at 0.86 kg a.i. ha^{-1}, F6875-7-1 at 1.26 kg a.i. ha^{-1}, and F6875-7-2 at 1.26 kg a.i. ha^{-1} provided excellent control (≥90%) of smooth crabgrass, while F6875-9 at 0.86 kg a.i. ha^{-1}, F6875-9 at 1.26 kg a.i. ha^{-1}, F6875-10 at 1.26 kg a.i. ha^{-1}, and Echelon GR on a fertilizer carrier (prodiamine + sulfentrazone) at 1.26 kg a.i. ha^{-1} provided good control (80 to 89%). Echelon (prodiamine + sulfentrazone) at 0.86 kg a.i. ha^{-1} provided fair control of smooth crabgrass after 5 months in bermudagrass. In both bermudagrass and tall fescue fields, F6875-7-1 and F6875-7-2 at 0.86 kg a.i. ha^{-1} provided poor control (<70%) of smooth crabgrass. F6875-7-1 and F6875-7-2 at 1.26 kg a.i. ha^{-1} in tall fescue controlled smooth crabgrass 73 to 86% after 5 months. Echelon on a fertilizer carrier at 0.86 and 1.26 kg a.i. ha^{-1} provided similar smooth crabgrass control in tall fescue after 5 months, and averaged 85%.

RECONSTRUCTING A PRIVET INVASION USING GENETIC, AGE, AND SPATIAL DATA. W. Zhao, A. Michel, and J. Cardina*, Ohio State University, Wooster, OH (73)

ABSTRACT

Ligustrum vulgare (Common privet or European privet) has escaped from cultivation as a colonial-era hedge species to become a serious invader of sensitive and unique habitats in eastern North America. As a case study of plant invasion into a natural area, we used genetic and dendrochronology data to reconstruct patterns of invasion of privet in and around a 325-acre park in northeast Ohio. We used two chloroplast DNA markers to characterize haplotypes within the park, and in natural areas throughout Ohio, New York, and Pennsylvania. Haplotype distribution across the three states suggests an east to west invasive pathway, with greater diversity at sample sites in New York and Pennsylvania than in Ohio. Maps of haplotypes for over 300 samples within the park show a mixed distribution pattern with no clear patterns of invasion trajectories inside the park. We also identified seven microsatellite markers and genotyped 140 samples; sibship analysis of those samples suggests constant invasion events in the park. Corresponding maps of annual growth rings and land cover types showed that the oldest patches were located at field-forest boundaries, possibly associated with animal activity. The local Gi* statistic identified old and young clusters of privet patches, which were likely influenced by landscape factors. The global Moran's I calculated in each year suggested a trend of clustering during invasion. The privet invasion pattern in the park was visualized by mapping the age class of sampled patches over time to show the formation of new patches during invasion period. The oldest patch formed 39 years ago followed by an approximate 20-year lag phase with few new invasions that were dispersed across the park landscape. After the lag phase there was a linear increase of about 20 new patches per year over the next 15 years, followed by a decline in new patch formation. Results reinforce the need for early detection a vigilant surveys to protect sensitive habitats from invasive plants like privet.

ASSESSING HAIRY VETCH SEED BANK PERSISTENCE AND POTENTIAL FOR MANAGEMENT. J. Wallace*, S. Mirsky, W. Curran, and B. Crockett, Penn State University, University Park, PA (74)

ABSTRACT

Hairy vetch (*Vicia villosa* Roth) is a legume cover crop that consistently outperforms (biomass, nitrogen fixation and winter hardiness) other legumes growing in temperate regions. Despite the benefits of this cover crop, its agronomic use, particularly among small grain producers, has been limited due to its potential weediness. Hairy vetch exhibits prolonged physical dormancy that results from a hard seed coat. In this study, we evaluated hairy vetch seedbank dynamics under field conditions as influenced by seed scarification, cultivar and soil depth.

We conducted field experiments in 2009-10 and 2010-11 growing seasons at the Rock Springs Agronomy Farm in Centre County, PA and the Beltsville Agricultural Research Center in Beltsville, MD. Experiments were designed as a split-split plot with four replications. Main plot effects consisted of two excavation dates (6 and 12 mo after burial) with seed burial depth (3 and 15 cm) as the sub-plot treatment. Within the sub-plots, four treatments consisting of hairy vetch cultivar (Albert-Lea and Groff-Auburn Early Cover) and seed scarification (scarified vs. non-scarified) were randomized. Each experimental unit consisted of 500 seeds that were placed in a 900 cm^2 aluminum mesh cage and buried using soil from fields without a history of hairy vetch use. Hairy vetch seedling emergence was monitored bi-weekly to monthly. Excavated seed cages were elutriated and recaptured seed was tested for viability using a tetrazolium test.

Seedling emergence patterns varied among hairy vetch varieties across years and study locations. In response to burial depth, seedling emergence varied across study location and year, but was similar across variety and scarification treatments. Deeper burial depth (15 cm) resulted in significantly lower emergence (2009: 35-46%; 2010: 7-35%) compared to the 3 cm depth (2009: 56-77%; 2010: 35-80%). Scarified seed, regardless of variety or burial depth, were not viable within 6 months of planting. In contrast, 1% to 3% of non-scarified seed were persistent in the seedbank 12 months after burial at the 3 and 15 cm burial depths, respectively.

In response to seed scarification, seedling emergence varied across varieties, burial depth and location. The net effect of seed scarification on total seedling emergence, in comparison to non-scarified seed, is likely a function of many interacting factors such as the proportion of hard seed in a variety, the impact of the scarification process on a seed cohort, and environmental factors that influence seed susceptibility to pathogens. Our results suggest that mechanical scarification prior to planting has the potential to eliminate hairy vetch seed bank persistence. Our results also indicate that deep tillage may be utilized to decrease the rate of hairy vetch seedling emergence in the subsequent crop rotation if hairy vetch completed its life cycle prior to termination or if there is concern for hard seed.

BIOMASS ALLOCATION OF TWO SWALLOWWORT SPECIES IN CONTRASTING COMPETITIVE ENVIRONMENTS. A. DiTommaso*, Z. Qin, M. Ho, C.L. Mohler, and K.M. Averill, Cornell University, Ithaca, NY (75)

ABSTRACT

Knowledge on how allocation patterns of plant biomass influencecompetitive ability may aid in elucidating invasiveness of alien plants. Water availability and intra- and inter-specific competition are considered important factors affecting biomass allocation patterns of the invasive perennial vines, *Vincetoxicum rossicum* (pale swallowwort) and *V. nigrum* (black swallowwort) Based on a glasshouse experiment, we investigated the growth of the two swallowworts and explored how shoot and root allocation of these herbaceous vines responded to four competitive regimes (shoot competition only, root competition only, full competition, no competition) at two levels of soil water availability and in the presence of conspecific or congeneric neighbors. Competition intensity was determined as the natural log of the relative biomass of a target plant grown with competition compared with growth in the absence of competition. We found that: (1) Under soil water deficit conditions, both *V. rossicum* and *V. nigrum* performed similarly. *Vincetoxicum nigrum* had a greater capacity to adjust to lower water availability by reducing plant height, number of nodes, and biomass accumulation as well as increasing the root–shoot ratio. *Vincetoxicum rossicum* experienced greater reductions in growth at lower water availabilities than *V. nigrum* although both species had high survival under low water conditions. However, *V. rossicum* performed better than *V. nigrum* under adequate soil water conditions as indicated by a significantly higher root–shoot ratio. (2) Root–shoot interactions had negative synergistic effects on target swallowwort species grown with conspecies but positive synergistic effects on target species grown with congeners. Positive interactions between root and shoot competition in mixed swallowwort communities tended to increase with high soil water availability. (3) Generally, *V. nigrum* maximized the capture of light and water resources by maximizing vegetative growth in the presence of neighbors, whereas *V. rossicum* plants responded by limiting above-ground growth and increasing allocation to root production, regardless of competitive environment and soil moisture level. Phenological differences between *V. rossicum* and *V. nigrum* may help to explain the various responses to the competition treatments and soil water availabilities experienced in this study. Given that the distributions of the two species may overlap in New York State and elsewhere in the future, management strategies that target mixed populations of these two *Vincetoxicum* species will be important and will be dependent on a thorough understanding of their responses to biotic and abiotic factors.

ON THE ECOLOGY OF TREE SPECIES AT 50 HOME SITES AT POWHATAN PARK, MONTVALE, NEW JERSEY. R. Stalter* and M.D. O'Donovan, St. Johns University, Queens, NY (76)

ABSTRACT

The objective of this study was to identify tree seedlings and saplings at 50 home sites at the Powhatan Park region of Montvale, New Jersey. A second objective was to map, and to determine ecological dominance of trees at these home sites as a potential seed source of colonizing trees for the larger region. Powhatan Park, (41.05, -74.05) includes Waverly Forest and Westmoreland Avenues and was developed as a summer resort complete with clubhouse and summer bungalows in the 1920's. Twenty-eight tree species have colonized the properties. White ash (*Fraxinus americana* L.), red oak (*Quercus rubra* Lam.), white mulberry (*Morus alba* L.), yellow poplar (*Liriodendron tulipifera* L.) were the most successful colonists, found at 42, 30, 28, and 28 percent of the home sites respectively. The number of colonizing tree species encountered at each home site ranged from 0 to 10. Most colonizing trees occupied property boundaries where their chance of removal (extirpation) was minimal. Property maintenance was the most likely cause of extirpation of colonizing trees at home sites.

The mode of seed dispersal of colonizing taxa follows. Birds were responsible for transporting seeds of black cherry (*Prunus serotina* Ehrh.), white mulberry, flowering dogwood (*Cornus florida* L.) and American holly (*Ilex opaca* L.) to the sites in their droppings. Wind dispersal brought seeds of American elm (*Ulmus americana* L.), maples (*Acer* spp.), white ash, and Norway spruce (*Picea abies* (Lunell) Rydb.) to the properties. Small mammals were responsible for caching seeds of oaks and hickories (*Carya* spp.). Humans were responsible for bringing potential seed sources to the area from plantings notably catalpa (*Catalpa speciosa* L.), callery pear (*Pyrus calleryana* Decne.), American holly, mimosa (*Albizia julibrissin* Durazz.) and Siberian elm (*Ulmus pumila* L.). Oaks were the dominant trees at the home sites; most exceed 100 years old.

SIMULATING INVASION ESTABLISHMENT AND SPREAD ACROSS A HETEROGENEOUS ENVIRONMENT WITH INVDER. D.R. Tekiela*, E. Dollete, M. Ho, and J.N. Barney, Virginia Tech, Blacksburg, VA (77)

ABSTRACT

Empirical studies have elucidated many components of the successful invasion of some our worst invasive plant species. Though these studies have furthered the wealth of knowledge of undesirable plant species, there has been no attempt to integrate the data from diverse studies into a meaningful framework to help understand invasions through the entire process of establishment to spread. Additionally, the landscape matrix in which invasions manifest are not homogenous, and present a diversity of invasible habitats. Understanding the process of invasion in its entirety is a critical component to determining the most effective land management strategies to mitigate establishment and spread of invasive species.

We created the Invasive Dispersal & Establishment Risk model (InvDER) as a spatially explicit stochastic simulation model that characterizes the invasion process across a heterogeneous landscape. By integrating geographic and land-use data, species demographic characteristics, species-specific dispersal kernels, and landscape-specific establishment probabilities this model is capable of more accurately characterizing the relative risk of establishment and spread of an invasive species. The InvDER model is capable of both explaining past invasions using historical introduction accounts (i.e., hindcasting), but can also be used to simulate future invasions over any timeframe (i.e., forecasting). Currently the model is being used to describe the invasion of *Miscanthus sinensis* from the initial point and time of introduction and forecasted into the future.

By manipulating each parameter (e.g., establishment probability, dispersal characteristics) within the model it is possible to run sensitivity analyses to better understand critical parameters required for successful invasion across a landscape. Additionally, this model can identify critical threshold values within parameters that determine whether invasion will be successful.

InvDER model results will lead to a more holistic understanding of the invasion process. By better understanding the most critical components of invasion, land managers will be better able to discourage the establishment and success of invasive species across heterogeneous landscapes.

ECOLOGICAL STUDIES ON DROUGHT TOLERANCE BY INVASIVE PLANT SPECIES. S.L. Young*, University of Nebraska-Lincoln, North Platte, NE (78)

ABSTRACT

What do extended periods of drought mean for managing invasive plant species in riparian areas and rangelands? This question will be addressed using two examples from research studies conducted with the past 3 years in Nebraska. Phragmites australis and Carduus nutans are two invasive plants impacting riparian areas and rangelands, respectively. Both were found to respond differently to limited water in greenhouse and field studies. Research on drought tolerant invasive plants will be important for understanding changes in plant communities and interactions between natives and invasive plant species.

JAPANESE KNOTWEED - WHY DETAILS ARE IMPORTANT. T.W. Curran*, retired, US Army Corps of Engineers, Huntington, WV (79)

ABSTRACT

What started as two field trials of herbicide by the landowner on *Fallopia* x *bohemica* in a rural West Virginia watershed ended in a draft theoretical framework of knotweed function. All of the required information was available in weed science. The first trial injected 0.65 grams of glyphosate as isopropylamine salt into 243 mature knotweed canes on 20 August 2012, which is also equivalent to 3 ml of retail packaged 18% solutions available to local farmers. Plants slowly died over a three week period. The Roundup PRO (reg. trademark of Monsanto technology LLC) label allowed 5 ml, equivalent a.i. of 3 grams glyphosate per cane of all three species of *Fallopia japonica* s. l. The next spring no mature canes survived, and the number of surviving small plants was about the same as the number of solid canes not treated. The second trial, on two very small infestations, of a combination of glyphosate with an additional 20% 2,4-D by weight of a.i. on 1 May 2013, by injection as on 20 August 2012, and 2% v/v as a foliar spray, resulted in a quick top kill and multiple new sprouts from underground rhizomes three weeks later. There was no significant difference in results for the two methods of application in the second trial. It is speculated that application timings, rates, or plant physiology may have caused these differential responses. However, Figure 3.2, Life cycle of a perennial weed… from Zimdahl, Fundamentals of Weed Science (2007) gives insight on the importance of the logistics of photosynthate recharge of the knotweed rhizome. Mass balance of water entering and leaving the rhizome must be maintained. Because photosynthesis is the engine of hydraulic translocation, photosynthesis must be maintained as long as possible to allow a systemic herbicide to translocate to the furthest reaches of the very extensive knotweed rhizome network. Powerful doses of injected herbicide at ground level and through foliar application terminated photosynthesis too quickly, which resulted in limited translocation. These trials translated scientific knowledge into cultural knowledge of use to small scale farmers and landowners in the watershed.

KNOTWEED - UNINTENDED CONSEQUENCES ON A GRAND SCALE. T.W. Curran*, retired, US Army Corps of Engineers, Huntington, WV (80)

ABSTRACT

Knotweed in West Virginia is primarily *Fallopia* x *bohemica*. In about 700 miles of travel in the summer of 2013, the author encountered only one large infestation of *Fallopia japonica*. So large a proportion of Bohemian knotweed cannot be explained by simple hybridization in the wild. Bohemian knotweed where observed appeared to be associated with riparian areas or with recent development. This implies propagation by floods, by contaminated heavy equipment or through invasion of disturbed land. Large population gains were associated with mowing. The Ohio River is heavily infested with Bohemian knotweed from the Pennsylvania border down to Wheeling, WV, and then infested to a lesser extent further downstream to the Kentucky border and beyond. Fill excavated from Ohio River or other floodplains could be contaminated with knotweed rhizome. Most knotweed infestations are on unoccupied land. Knotweed is just beginning to invade agricultural land, but is not yet an urban or suburban pest. *Fallopia japonica* has been observed in England to clog drainage lines, punch through brick walls, destroy asphalt paving, and invade houses. The USA has *Fallopia* x *bohemica*, a more invasive and more difficult species. The scope of the knotweed problem is estimated from interviews, secondary sources of data, and similar data from Great Britain and the Czech Republic. Because the knotweed invasion in the USA is qualitatively and quantitatively different from that in Europe, the presentation cannot give firm figures, but only provide economic and environmental implications for the state of West Virginia. Lastly, the presentation gives a short list of management options from current weed science, including advice on use of ALS inhibitors and resistance prevention, and a long-term control strategy from Cornwall, England, which has been very active in knotweed control, and has cultural and economic similarities with West Virginia.

PRELIMINARY RESULTS: CHEMICAL CONTROL OF WILD CHERVIL ON ROADSIDES. B.F. McDonnell*, R.J. Hodge, and D.C. Gatch, National Park Service, Bushkill, PA (81)

ABSTRACT

Wild chervil (*Anthriscus sylvestris*) is becoming a common roadside weed in the Northeast, is a Class B noxious weed in Washington State and is prohibited in Massachusetts. Wild chervil is not known to be listed as a controlled weed on any herbicide labels, and control methods have not been well documented. The following six herbicide rates, plus a control, were replicated three times in 3 by 6 meter plots: 1.96 kg a.e./ha glyphosate IPA, 0.84 kg a.e./ha imazapyr, 0.021 kg/ha metsulfuron methyl, 2.24 kg a.e./ha triclopyr BEE, 0.11 kg a.e./ha imazapic, and 0.12 kg a.e./ha aminopyralid. All chemical mixtures were applied at approximately 40 gallons per acre, with water, a non-ionic surfactant and blue colorant. The control plots received water, non-ionic surfactant and blue colorant. Applications were performed on April 23, 2013, and assessed on September 17, 2013. Results from 147 DAT: Metsulfuron methyl and imazapyr, near 100% control; glyphosate IPA 95% control; imazpic 91% control; triclopyr BEE 72% control, aminopyralid 57% control.

SURVEY, DECISION MAKING, AND MANAGEMENT RELATED TO MONOECIOUS HYDRILLA IN A NATURAL CAROLINA BAY LAKE, LAKE WACCAMAW. R.J. Richardson*, North Carolina State University, Raleigh, NC (82)

ABSTRACT

Lake Waccamaw is a natural, shallow lake in southeastern North Carolina with several endemic species. Monoecious hydrilla was first confirmed in Lake Waccamaw in early fall 2012. A technical advisory committee (TAC) was formed and a lake wide vegetation survey conducted in late fall 2012. A hybrid SONAR/point intercept survey was used to quantify aquatic vegetation in the lake. The survey included 362 point intercept samples collected by rake toss and 83 miles of continuous SONAR data. SONAR and point intercept data were merged to create vegetation maps and calculate plant distribution area. The survey indicated that the 3,617 ha lake contained 246ha of hydrilla and 1,456 ha of native vegetation. With maximum depth of the lake being 3.4m, it was estimated that hydrilla could cover 100% of the lake area and that 2/3 of the lake would be infested within 5 years. The TAC conducted a risk analysis of possible management efforts including lake drawdown, triploid grass carp, mechanical harvesting, and herbicide treatment. Targeted herbicide treated was selected by the TAC as the initial control strategy due to a combination of higher efficacy and lower risk than other options. Fluridone was applied to a treatment are of approximately 366ha in 2013. Point intercept surveys were conducted several times during the 2013 growing season with very few hydrilla fragments recovered. Those found were typically in poor condition and exhibiting treatment symptomology. The high level of hydrilla control obtained was likely due to a combination of the herbicide treatment and high water levels from abnormal precipitation levels. A second full lake hybrid SONAR/point intercept survey is planned for October 2013 to fully quantify vegetation changes.

2014 ACTIVITY REPORT FOR THE MASSACHUSETTS INVASIVE PLANT ADVISORY GROUP. R. Prostak*, University of Massachusetts-Amherst, Amherst, MA (83)

ABSTRACT

The Massachusetts Invasive Plant Advisory Group [MIPAG] is a voluntary organization of individuals representing state and federal agencies, academic institutions, land conservation organizations, as well as nursery industry professionals, all of whom are concerned with the conservation of the Massachusetts landscape. First assembled in 1995, MIPAG was charged by the Massachusetts Executive Office of Environmental affairs to provide recommendations regarding which plants should be deemed invasive, and what steps should be taken to manage these species. MIPAG, with the assistance and direction of Leslie Mehrhoff of the University of Connecticut developed and adopted definitions, created a list of plant to be evaluated, and develop a set of biologically based criteria upon which to evaluate objectively plants suspected to be invasive in the Commonwealth. A total of 85 species were evaluated and identified as *Invasive*, *Likely Invasive*, or *Potentially Invasive*. Species evaluated in which sufficient information or evidence is currently lacking for an adequate evaluation were designated as *plants not meeting criteria (do not list at this time)*. An annotated list of all species evaluated was developed. Upon the completion of plant species evaluations, MIPAG developed and outlined invasive-plant-management objectives for the Commonwealth in *Strategic Recommendations for Managing Invasive Plants in Massachusetts*.

Recent MIPAG activities

1. Massachusetts Invasive Plant Species: Early Detection Priorities

Consistent with its Strategic Plan to implement an early detection and rapid response plan for the Commonwealth, a MIPAG develop a list of Early Detection (ED) plant species. Tasks included establishment of criteria for selecting plants and compiling the Early Detection species list.

2. Guidance for the Effective Management of Invasive Plants

MIPAG developed this document in order to provide step-by-step guidance for landowners and land managers in developing specific plans for the effective management of invasive plants. The guidance is intended to work at any scale, from a small parcel of land to statewide. The main focus of the document is invasive plant control for ecological health, although other goals of invasive plant management (such as human health and aesthetics) are compatible with this methodology. The document also provides a management case study as well as links to other relevant resources

Background information about MIPAG and selected MIPAG publications can be found at www.mipag.net

FURTHER INVESTIGATIONS INTO HERBICIDE SELECTION AND APPLICATION TIMING TO SUPPRESS JAPANESE STILTGRASS. A.E. Gover*, J.M. Johnson, D.A. Despot, and J.C. Sellmer, Penn State University, University Park, PA (84)

ABSTRACT

Current operational herbicide regimens to suppress the exotic, annual Japanese stiltgrass (*Microstegium vimineum* (Trin.) A. Camus *var. imberbe* (Nees) Honda, MCGVM) include preemergents, postemergent graminicides, and spot applications of non-selective herbicides. This experiment evaluated the selective broadleaf herbicides aminocyclopyrachlor and aminopyralid, and evaluated application timing impact on suppression from aminocyclopyrachlor, quizalofop, or a low rate of glyphosate. The June 11, 2013 application included aminocyclopyrachlor at 0.075, 0.10, or 0.13 kg/ha, alone or as a pre-mix in combination with chlorsulfuron at the corresponding rates of 0.032, 0.039, or 0.053 kg/ha, respectively; aminopyralid at 0.12 kg/ha, quizalofop at 0.031 kg/ha, and glyphosate at 0.21 kg/ha. Application timing was evaluated with a July 10 application of aminocyclopyrachlor at 0.075 kg/ha, quizalofop or glyphosate, in addition to aminocyclopyrachlor plus triclopyr at 0.075 plus 1.7 kg/ha to determine if triclopyr, a likely tank-mix partner, antagonized aminocyclopyrachlor activity. All treatments were applied to 1.8 by 4.6 m plots arranged in a randomized complete block with four replications along a highly disturbed, forested roadside in Bald Eagle State Park, Howard, PA. Data collected on June 11, July 10, August 16, and September 17 included percent total vegetative cover and percent MCGVM cover. A floristic census for each plot was taken at the June, July, and August ratings. Percent values and species counts were subjected to analysis of variance, and means compared using Fisher's Protected L.S.D. Pre-treatment total vegetative cover and MCGVM cover were not different between treatments. Total vegetative and MCGVM cover in the untreated plots averaged 44 and 33 percent in June, and 56 and 50 percent in September, respectively. All herbicide treatments significantly reduced MCGVM cover by September (0 to 5 percent), and there was not a significant timing effect or herbicide by timing interaction on MCGVM reduction for the low rate of aminocyclopyrachlor, quizalofop, or glyphosate. Pre-treatment, 40 species (excluding MCGVM) were recorded, averaging 4.0 to 6.8 per assigned treatment. In August, there were 44 species (excluding MCGVM), ranging from 1.8 to 4.5 per treatment, with 3.8 in the untreated plots. Treatment effect was significant for species number in September, and change in species number from pre-treatment to September, but there was no apparent trend related to active ingredient or application rate.

COMPARISON OF HERBICIDE MIXTURES FOR CONTROL OF MORROW'S HONEYSUCKLE. J.M. Johnson*, D.A. Despot, and J.C. Sellmer, Penn State University, University Park, PA (85)

ABSTRACT

Exotic shrub honeysuckle, which often includes a mix of Tatarian, amur, and Morrow's honeysuckle species, has become a commonplace problem along the roads of Pennsylvania. As with other woody species invading transit corridors, exotic shrub honeysuckle limits sight-distance and interferes with highway maintenance activities necessitating their control. We have observed that herbicide tank mixes commonly used to manage other encroaching tree and shrub species with foliar sprays are often ineffective at controlling the exotic shrub honeysuckles. Glyphosate has been effective, however, it readily damages the understory. The goal of this experiment was to identify grass safe herbicide combinations that were effective at controlling an exotic shrub honeysuckle, specifically Morrow's honeysuckle. The experiment compared both newer (aminocyclopyrachlor and aminopyralid) and older chemistry with selectivity to grasses. Ten herbicide treatments and treatment combinations were tested including: 269.5 g aminocyclopyrachlor plus 14 g metsulfuron/ha; 269.5 g aminocyclopyrachlor plus 24.5 g metsulfuron/ha; 1683 g triclopyr/ha plus 134.7 g aminocyclopyrachlor and 28.1 g metsulfuron/ha; 1683 g triclopyr/ha plus 68.8 g aminocyclopyrachlor and 21.8 g metsulfuron/ha; 1683 g triclopyr/ha plus 123.4 g aminopyralid and 22.4 g metsulfuron/ha; 1683 g triclopyr/ha plus 2131g 2,4-dichlorophenoxyacetic acid/ha; 1683 g triclopyr/ha plus 3365 g glyphosate/ha; 841 g triclopyr/ha plus 1122 g dicamba/ha and 123.4 g aminopyralid/ha; 3365 g glyphosate/ha alone; 2356 g 2,4-dichlorophenoxyacetic acid and 897.5 g triclopyr/ha plus 21.1 g metsulfuron/ha; and an untreated check. All herbicides included a non-ionic surfactant at 0.25 percent v/v.

Treatments containing either glyphosate or 2,4-dichlorophenoxyacetic acid produced the most significant impact on Morrow's honeysuckle nearly one year after treatment, YAT, with values of 83 to 97 percent control. The triclopyr plus 2,4-dichlorophenoxyacetic acid treatment demonstrated a promising level of control of the Morrow's honeysuckle (89%) while preserving the grass understory. All other treatments resulted in low to moderate control from 19 to 51 percent. Injury to grasses was greatest with treatments containing 269.5 g aminocyclopyrachlor plus 14 or 24.5 g metsulfuron/ha, 134.7 g aminocyclopyrachlor plus 28.1 g metsulfuron/ha, or glyphosate combinations whereas forbs were equally damaged by all treatments. Further testing of 2,4-dichlorophenoxyacetic acid in progressive rates and combinations within the label guidelines is necessary to assure efficacy and safety prior to making a clear recommendation for its use in control of the troublesome exotic honeysuckles.

MEASURING THE ECOLOGICAL IMPACT OF INVASIVE PLANT MANAGEMENT IN SUCCESSIONAL PLANT COMMUNITIES ALONG CONNECTICUT'S COAST. C.H. Douglass*, Trinity College, Hartford, CT (86)

ABSTRACT

Invasive plant species have become ubiquitous across the landscape of the northeastern U.S., but are arguably most aggressive and problematic in successionally transitional habitats such as riparian zones and forest or old field edges. In these intermediate habitats variable edaphic and environmental conditions appear to favor the establishment of assemblages of more highly adaptable invasive plant species. The development of dominant assemblages of invasive vines and shrubs in particular in these transitional habitats seems to accelerate and alter natural successional trajectories. On a local community scale this project aims to quantify the ecological impacts of managing assemblages of woody invasive plants in transitional habitats using adaptive management strategies. On a broader site scale we seek to characterize whether and how these management efforts affect the trajectory of the successional transition of the intermediate habitat.

A pilot study was initiated in summer 2013 at a privately owned nature preserve in coastal Connecticut containing an early successional grassland and an area of rare shrubland. Between these habitats is a relic stone wall overgrown with an assemblage of invasives dominated by black swallow-wort (*Cynanchum nigrum* (L.) Pers., non Cav.), Amur peppervine (*Ampelopsis brevipedunculata* (Maxim.) Trautv.), and Oriental bittersweet (*Celastrus orbiculatus* Thunb.). Preliminary treatments occurred in early fall 2013 with a foliar application of 5% glyphosate + 1% metsulfuron methyl made to the target area. Nested 5 X 5 m vegetation sampling plots were randomly located in the transitional area prior to these treatments and plant species diversity, density and productivity measured. Transects were established crossing the entire site to quantify structure and density of the understory and overstory plant communities, and were also used for the systematic collection of soil samples. The ultimate management objective at this specific site is to promote the establishment of native species that provide structurally diverse habitat for threatened bird species, and so the relative success of the habitat management efforts will also in part be measured with metrics of avian species diversity and habitat usage.

Preliminary pre-treatment data suggested that overstory canopy density and light interception stratified invasive species distributions across the relatively narrow area. Amur peppervine and reed canary grass (*Phalaris arundinacea* L.) were more common and productive under full sun conditions, while black swallow-wort and Japanese honeysuckle appeared more vigorous in partial shade. Analysis of historical and contemporary bird banding records revealed that establishment of the woody invasive assemblage in the intermediate habitat mirrored a significant decline in species diversity. While post-treatment community data and additional broader scale sampling for edaphic factors remains to be completed, early results from this pilot study suggest that our study protocol will allow us to successfully monitor both local community and larger-scale site responses to invasive plant species management in sucessionally transitional habitats.

USING RESEARCH DRIVEN CRITERIA FOR DEVELOPMENT OF A LONG TERM AQUATIC PLANT MANAGEMENT PLAN: REVITALIZATION AFTER 15 YEARS OF MANAGEMENT ON LAKE GASTON. B. Hartis* and R.J. Richardson, North Carolina State University, Raleigh, NC (87)

ABSTRACT

Management of monecious hydrilla on Lake Gaston has remained a concern and financial burden for the States of Virginia and North Carolina for nearly three decades. In 2005, the entities charged with developing a plan for the Lake set an overall goal to develop and maintain a healthy lake ecosystem based on a diverse plant community dominated by native species. This included the goal of reducing the overall hydrilla acreage to less than 300 by 2012. Despite intense management through herbicides and grass carp stockings, investigative research and revegetation efforts, hydrilla levels have remained around 1500 acres for nearly a decade. Funding limitations and inconsistencies have led to “hop-scotch” treatment of areas further complicating the dynamic issue of lake-wide hydrilla management. Furthermore, a lack of communication between funding sources and the lake management group as well as State funding limitations have led to a gradual decrease in annual funding for the past few years. In spring of 2013, redevelopment of a long term management plan for the lake was undertaken to address these issues while also introducing a research driven model for treatment site selection. The plan was designed primarily to address not only hydrilla biomass, but to take into account tuber bank regeneration in a Lake with a long history of hydrilla infestation. Treatment site selection was based on a combination of decision tree modeling and expert opinion. The plan addresses long term lake management through selection of Long Term Treatment Areas (LTTAs) while also taking into account short term need for treatment at Priority Treatment Areas (APTAs) on an annual basis. Sites selected as LTTAs was highly reliant upon recent previous treatments, county distribution , size, and recreation and tourism value. The overall goal of the plan is to reduce existing hydrilla acreage to maintenance levels which would reduce treatment expenses and realize the initial goals of the lake management group set in 2005. This plan should provide a road map for hydrilla reduction within the lake over the long term while also providing measureable results for funding agencies year to year.

MODELING THE POTENTIAL GEOGRAPHIC DISTRIBUTION OF HYDRILLA IN NORTH AMERICA AND BEYOND. B. Hartis* and R.J. Richardson, North Carolina State University, Raleigh, NC (88)

ABSTRACT

Hydrilla (*Hydrilla verticillata*) is a highly invasive submersed aquatic weed. In North America, the troublesome invader has rapidly spread as far North as Maine and as far west as Washington State since the original introduction into Florida during the 1950s. The establishment of Hydrilla in such cool environments has recently raised questions about the plants ability to spread even further north on the continent. Recent discoveries of monoecious hydrilla in Lake Cayuga, NY, the Erie Canal, and the Ohio River have further highlighted the need for addressing the potential establishment of this submersed macrophyte. In our study, we attempt to model the establishment potential of hydrilla using a modified version of the climate envelope method, based on the plants known geographic distribution in the Northern Hemisphere, particularly along its northernmost limits. The objectives of this study were to model the potential establishment of hydrilla in the United States and Canada based on known occurrence sites worldwide and to determine if the establishment potential of hydrilla in various states/ provinces differs based on the size of water bodes present. Our model predicted that nearly the entire U.S. including various provinces in Canada were at high risk. More importantly, states and provinces that may have been thought to be far beyond the previously modeled northern limits of hydrilla's establishment threshold were well within climate suitable for establishment. Furthermore, differences in risk were apparent between water bodies of varying size. Results from this study should inform researchers and resource managers alike of the establishment potential of hydrilla along norther tier states and provinces in North America. Water body types identified within each state of higher risk should aid managers in appropriately allocating prevention, monitoring and management resources to prevent the spread of hydrilla into previously uninfested water bodies.

CORN RESPONSE TO SOIL-APPLIED ALS-INHIBITING HERBICIDES. M.J. VanGessel*, Q. Johnson, and B. Scott, University of Delaware, Georgetown, DE (89)

ABSTRACT

Many of the newer soil-applied herbicides that contain acetolactate synthase inhibiting (ALS) (or Group 2) herbicides or isoxaflutole (Group 27) have not been compared for relative crop safety when used on coarse-textured soils. Furthermore, most weed control trials are not designed to evaluate the impact of herbicide injury on corn yield. This trial compared relative crop safety of soil-applied herbicides containing ALS-inhibiting herbicides or isoxaflutole, and the resulting impact on yield. In addition, corn hybrid sensitivity to these herbicide families was also evaluated.

The trial was initiated in 2012 and repeated in 2013. The study was a strip plot with herbicides as the whole plot and corn hybrid as the sub-plot. Herbicide treatments were preemergence applications of Hornet WDG Premix, Basis Blend Premix, Capreno Premix, Prequel Premix, Corvus Premix, Balance Flexx, Instigate Premix, Instigate Premix followed by postemergence (POST) application of Realm Q Premix, Realm Q Premix applied POST and an untreated plot. Herbicides were applied at twice the labeled rate. Plots were maintained weed free. .

Field corn was planted April 19, 2012 and May 6, 2013 and herbicides applied within 24 hours of planting. In 2012 Hubner hybrids H5707, H5753, and 5505VT3P were planted and in 2013 H6652, H4744, H5555 and H6179 were used. Hybrids were selected to provide a range of sensitivity to Groups 2 and 27 herbicides.

In 2012, there was not a significant herbicide by hybrid interaction for corn injury. There were significant treatment differences, but no treatment resulted in >13% stunting at 3 and 4 weeks after planting (3 and 4 WAP). Balance Flexx and Capreno exhibited the least amount of stunting. There was no treatment difference in corn yield.

In 2013, different corn hybrids were used, and the site received 1.2 inch of rainfall within one week of planting. For the first 4 WAP there was a hybrid by herbicide interaction. As the season progressed, the main factor of hybrid was significant but the interaction was not significant, with H5555 exhibiting the most injury. At 4 WAP, Balance Flexx treatments were not exhibiting any stunting, Corvus treatments were rated at 29% stunting, and all other treatments were >40%. By 8 weeks after planting, all treatments were rated less than 10%, except Instigate followed by Realm Q which was 18%. No significant hybrid by herbicide interaction was detected for yield. Plots treated with Balance, Hornet, and Corvus treatments had similar yields to the untreated check. Capreno, Instigate, Basis Blend, Realm Q, and Prequel all had similar yields (averaging 14% lower yield than the untreated check), with Instigate followed by Realm Q as the lowest yield at 26% yield loss.

Under unfavorable conditions, (coarse textured soils, overlaps, lower temperatures, and high rainfall prior to corn emergence) corn injury from some soil-applied herbicides can be severe and yield loss can result. However, yield loss may not be as severe as early-season injury indicates.

DANDELION MANAGEMENT IN ZONE-TILLAGE GLYPHOSATE-RESISTANT CORN AND SOYBEANS. R.R. Hahn*, R.J. Richtmyer III, and J.M. Orlowski, Cornell University, Ithaca, NY (90)

ABSTRACT

A rotation experiment was established near Aurora, NY in 2010 to determine the value of reduced rates of residual herbicides in preventing dandelion (*Taraxacum officinale* Weber in Wiggers) encroachment into zone-tillage corn (*Zea mays* L.) and soybeans (*Glycine max* Merrill). The field was fall plowed in 2009 to eliminate established dandelions. Four-year rotations established as main plots (12 rows by 300 ft) with five replications were: 1) continuous corn with residual herbicides, 2) corn, soybeans, corn, soybeans with residual herbicides in both crops, 3) soybeans, corn, soybeans, corn with residual herbicides in both crops, 4) corn, soybeans, corn, soybeans with residual herbicides in corn only, and 5) soybeans, corn, soybeans, corn with residual herbicides in corn only. Sub-plots (12 rows by 75 ft) within crops received no residual herbicides, one-half, two-thirds, or full labeled rates of residual herbicides. Corn sub-plots were treated early postemergence (EPOST) with 0.77 lb ae/A of glyphosate alone or tank-mixed with 1.23, 1.64, or 2.46 lb ai/A of a premix of *S*-metolachlor, atrazine, and mesotrione (Lumax). Soybean main plots were split into six row strips so two residual programs could be compared. In one strip, sub-plots received no residual or a premix of metribuzin and chlorimuron-ethyl (Canopy) at 0.84, 1.13, or 1.69 oz ai/A preemergence (PRE) followed by 0.77 lb/A of glyphosate mid-postemergence (MPOST). In the other strip, sub-plots received no residual or a premix of chlorimuron-ethyl, flumiclorac, and thifensulfuron-methyl (Enlite) at 0.67, 0.89, or 1.34 oz ai/A PRE followed by 0.77 lb/A of glyphosate alone or tank-mixed with a premix of chlorimuron-ethyl and thifensulfuron-methyl (Synchrony XP) at 0.05, 0.07, or 0.11 oz ai/A MPOST. Glyphosate-resistant corn 'DKC 4272' and soybeans 'AG 2130' were used. Starting in 2011, dandelion counts were made in a 7.5 by 75 ft area in the center of all sub-plots each spring prior to zone building and planting. When averaged across all rotations in spring 2013, there were 732 dandelions/1,000 sq ft in sub-plots that have had EPOST or MPOST applications of glyphosate only. After three seasons of continuous corn, dandelions averaged 624, 25, 21, and 12/1,000 sq ft in sub-plots receiving none, one-half, two-thirds, or full rates of the residual corn pre-mix respectively. In rotations where residual herbicides were used in both crops, dandelions averaged 769, 97, 53, and 21/1,000 sq ft for sub-plots receiving none, one-half, two-thirds, or full rates of residual herbicides respectively. In rotations with residual herbicides in corn only, 2013 counts were higher in the cycle that began with soybeans (no residual in the first year) than in the cycle that began with corn. In the rotation that began with soybeans, dandelions averaged 791, 608, 749, and 660/1,000 sq ft for sub-plots receiving none, one-half, two-thirds, or full rates of the residual corn program respectively. Counts for the cycle that began with corn were 708, 312, 243, and 270/1,000 sq ft for sub-plots receiving none, one-half, two-thirds, and full rates of the residual corn program respectively.

MANAGING A FEED GRAIN INITIATIVE: POINTERS AND PITFALLS. W.J. Everman*, R. Weisz, and R. Heiniger, North Carolina State University, Raleigh, NC (92)

ABSTRACT

The North Carolina livestock industry is not only important to our state but also to the nation. We are *second* among states in cash receipts from all poultry and eggs, *second* among states in pig inventory and pig crop, and *sixth* among states in cash receipts from livestock, dairy and poultry. The problem is that North Carolina was eighteenth among states in wheat and soybean production, and nineteenth among states in corn production in 2010. Sorghum production was negligible. In other words, we are a grain-deficit state and the deficit is about 2/3 of the required feed supply. The deficit varies year to year but is usually at least 200 million bushels depending on NC production, exports, and other local uses. We have been importing feed grain by rail from the Midwest for many years. The current crisis has come about because the rates charged by the railroads to bring the feed supply to the North Carolina livestock producers has been rising 7-8% per year and is now around 20% of the July corn futures price. The railroads are also requiring large orders to earn their business as other industries such as the military are in competition for their services and there is a limited supply of rail services. The livestock producers are on the verge of not being able to do business profitably in North Carolina.

The meat and poultry production and processing industries are major employers, sources of income, and tax base in rural North Carolina. Loss of livestock and poultry production reduces processing requirements so the combined economic impacts are large. Large negative economic impacts have been felt in communities where processing plants closed after poultry production was reduced. If this continues, rural North Carolina is in for seriously troubled times.

North Carolina's livestock producers, government, university personnel, and others have seen an opportunity that would be a win-win situation for feed grain and livestock producers. If North Carolina can expand its grain production and profitably sell it to the livestock producers at a price that is less than the total cost of purchasing and transporting the grain from the Midwest, then both industries will benefit, as well as the citizens of rural North Carolina.

Grain sorghum has been grown off and on in North Carolina. But, until now, there has not been much of a market. Grain sorghum has the advantage of being able to tolerate drought, so it is a more consistent crop with potentially lower production costs than corn. Murphy Brown is now willing to make a market for grain sorghum produced in North Carolina. They are willing to offer a production contract for grain sorghum at 95% of the corn price. This is an opportunity for feed grain growers to *potentially* expand their grain production, increase their profit, and to grow a crop that produces well on marginal corn land. Grain sorghum also is a potential rotation crop that could help control resistant Palmer amaranth on Roundup Ready corn, soybean, and cotton acres.

ENLIST E3™ SOYBEAN TOLERANCE AND WEED CONTROL WITH PRE FOLLOWED BY POST HERBICIDE PROGRAMS. B.D. Olson*, D.C. Ruen, J.M. Ellis, D.M. Simpson, and J.A. Huff, Dow AgroSciences LLC, Geneva, NY (93)

ABSTRACT

The Enlist™ Weed Control system is being developed in multiple crops and includes Enlist™ soybean, Enlist E3™ soybean.. Enlist is a weed control system composed of new herbicide-tolerant traits and a new herbicide solution, Enlist Duo™ herbicide. Regulatory approvals are pending for components of the Enlist system. Enlist soybean when stacked with glyphosate-tolerant traits, such as Roundup Ready 2 Yield, and Enlist E3 soybean will provide tolerance to glyphosate, glufsosinate and 2,4-D. Integrating multiple modes of action herbicides into a preemergence followed by postemergence weed control program provides consistent, highly effective control and helps prevent the onset of herbicide- resistant weeds.

A total of 22 studies were conducted in 2013 in the U.S. to evaluate the weed control delivered by a systems approach composed of PRE followed by POST herbicide applications in Enlist E3 soybeans. PRE foundation treatments consisted of cloransulam + sulfentrazone, flumioxazin + cloransulam, flumioxazin + chlorimuron ethyl or *S*-metolachlor + fomesafen herbicide products . Postemergence treatments were Enlist Duo (2,4-D choline + glyphosate DMA) at 1640 and 2185 g ae/ha, glufosinate at 542 g ae/ha, 2,4-D choline + glufosinate at 800 + 542 and 1065 + 542 g ae/ha, and glyphosate at 1120 g ae/ha applied approximately 30 days after planting.

PRE applications of cloransulam + sulfentrazone or flumioxazin + cloransulam followed by Enlist Duo at 1640 or 2185 g ae/ha or 2,4-D choline + glufosinate at 800 +542 or 1065 + 542 g ae/ha provided greater than 95% control of CHEAL, IPOSS and AMAPA and glyphosate resistant AMATA, AMBEL and AMBTR. Enlist Duo and 2,4-D + glufosinate treatments resulted in less than 3% visual soybean injury at 14 days after application.

Enlist E3™ soybeans are being jointly developed by Dow AgroSciences and MS Technologies.

ENGENIA™ HERBICIDE AS A WEED MANAGEMENT TOOL. J. O'Barr*, W.E. Thomas, J. Frihauf, S. Bowe, and L. Bozeman, BASF Corporation, Hummelstown, PA (94)

ABSTRACT

New weed control options are needed to manage herbicide resistant weeds that are limiting control tactics and in some areas cropping options. Dicamba tolerant soybean and cotton will enable the use of dicamba to manage these problematic weeds with an additional herbicide mechanism-of-action. In addition to being a new control tactic, these dicamba tolerant cropping systems will allow for application of dicamba as a preplant burndown without a planting interval and postemergence over the top of the crop. Engenia™ herbicide, currently not registered by the US EPA, is an advanced formulation based on the proprietary BAPMA (N, N-Bis-(aminopropyl) methylamine) dicamba that reduces potential secondary loss more than the improvement achieved with Clarity® herbicide. In addition to formulation innovation, a comprehensive stewardship strategy will be implemented to focus on weed management and effective control, weed resistance management, and maximizing on-target application.

Engenia herbicide should be used as a complementary tool in a grower's weed control program where it is integrated into a comprehensive strategy that includes cultural, mechanical, and chemical control. A robust herbicide program uses sequential and/or tank mixtures of herbicides that have multiple effective sites of action on a single weed. Likewise, Engenia should complement current programs to add an additional effective site of action for broadleaf weed control. Over several years of testing, the most effective soybean weed control programs have utilized preemergence followed by postemergence applications like Optill® PRO followed Engenia plus glyphosate.

In order to maximize on-target deposition, many parameters related to equipment setup and environmental conditions should be considered. Nozzle selection offers the opportunity to dramatically reduce the potential for spray drift. Research shows that venturi-type nozzle technology can significantly reduce drift potential. Other application parameters that should be considered include travel speed, boom height, application volume, use of a deposition aid, and proximity to sensitive crops. BASF has initiated the 'On Target Spray Academy' training series to educate applicators on best application practices. The combination of Engenia and dicamba tolerant crops plus a stewardship strategy will provide growers with an effective system to control increasingly difficult and herbicide-resistant broadleaf weeds.

CURRENT AND FUTURE STEWARDSHIP TRAINING FOR ENGENIA™ HERBICIDE.

L. Bozeman*, R.E. Wolf, S. Wilson, and D. Pepitone, BASF Corporation, Research Triangle Park, NC (95)

ABSTRACT

A diverse approach to herbicide use along with targeted spray application, are key stewardship components of an effective weed management program. The goal of herbicide application is to remove weeds that may compete with the crop and reduce crop yield. Ineffective weed control may result from applications that do not use a diverse approach to herbicide selection. A new tool currently in development is Engenia™ herbicide, an advanced formulation of dicamba for dicamba-tolerant soybeans and cotton. The formulation further improves on-target application through reduced secondary loss potential. A herbicide application that moves off target may cause unintended effects on non-target plant species that are contacted. It is important that applicator training programs be made available to provide guidance on herbicide systems and proper application to mitigate off-target sensitive plant damage and to maximize product performance. Training efforts underway and in development will be discussed.

RED CLOVER VS. HAIRY VETCH: WEED SUPPRESSION, CROP PERFORMANCE, AND PROFITABILITY IN A NO-TILL WHEAT-CORN ROTATION. E.M. Snyder*, W.S. Curran, H.D. Karsten, and G.M. Malcolm, The Pennsylvania State University, Altoona, PA (96)

ABSTRACT

Red clover (*Trifolium pretense* L.) and hairy vetch (*Vicia villosa* Roth) are leguminous cover crops ('green manure' hereafter) that are widely studied for their nitrogen-fixing contributions.This study evaluated weed control, crop yields, herbicide use, and net returns to management using each of these green manures in a winter wheat (*Triticum aestivum* L.)- green manure- corn silage (*Zea mays* L.) cropping sequence over two growing seasons in 2011 and 2012. Our results showed that weed infestation in wheat and corn did not differ between the green manure systems, but a second separate experiment (2012 only) showed more weed biomass before corn planting where there was no green manure and higher weed biomass late summer in hairy vetch-triticale (x *Triticosecale)* managed corn compared with no cover crop. In 2012 only, wheat and corn yields were greater in the red clover sequence by 182 and 5751 kg ha^{-1}, respectively compared to hairy vetch-triticale. Corn yield was likely reduced in the hairy vetch-triticale green manure due to corn population differences that resulted from early season pest pressure that was less of a problem in the red clover-corn system. Corn after hairy vetch-triticale also had lower plant population, but higher yield compared with no-cover, and corn yield was also greater following red clover compared with corn following no cover crop. The use of red clover compared to hairy vetch eliminated the need for two herbicide applications and used 40% less herbicide active ingredient. Primarily because of red clover forage value (a fall cutting of red clover was harvested), the red clover system resulted in greater net returns to management by $1,399.67 ha^{-1} over the hairy vetch-triticale system. This experiment demonstrated that establishing red clover between winter wheat and corn offered several advantages over hairy vetch plus triticale in the sequence. However, including either of these green manure crops before corn will eliminate the need for some nitrogen fertilizer and result in better corn yields as compared to corn preceded by no cover crop.

MESOTRIONE AND PRODIAMINE COMBINATIONS FOR SMOOTH CRABGRASS CONTROL IN TALL FESCUE. J. Yu* and P. McCullough, University of Georgia, Griffin, GA (98)

ABSTRACT

Mesotrione effectively controls 1 to 2 tiller smooth crabgrass in spring, but subsequent germination of crabgrass in summer may warrant tank-mixtures with preemergence herbicides for season-long control. Field experiments were conducted from April to September 2013 in Griffin, GA to evaluate mesotrione and prodiamine combinations, compared to dithiopyr and prodiamine alone, for postemergence and residual control of smooth crabgrass in tall fescue. Treatments were applied to an irrigated 'K-31' tall fescue lawn maintained at 7.5-cm mowing height with clippings returned. Mesotrione plus prodiamine treatments were made with a pre-packaged formulated mixture (4SC) at 0.56 kg a.i. ha^{-1} at three timings including: preemergence followed by (FB) 8 weeks after treatment (WAT), multi-leaf postemergence FB 4 WAT, or 1 to 2-tillered postemergence FB 2 WAT. Prodiamine alone was applied at the aforementioned preemergence program at 0.42 FB 0.42 kg a.i. ha^{-1}, or sequentially at multi-leaf postemergence FB 4 WAT. Dithiopyr was also applied sequentially at 0.28 kg a.i. ha^{-1} at the aforementioned preemergence program, or singly at 0.56 kg a.i. ha^{-1} at the two postemergence timings. Smooth crabgrass cover increased over time in the nontreated plots, and averaged 54% cover by September. All mesotrione + prodiamine treatments controlled smooth crabgrass >97% within 3 weeks after applications, and control was >90% on September 8. Sequential prodiamine treatments alone controlled smooth crabgrass 71% and 61% when applied at the preemergence and early postemergence timing, respectively. All dithiopyr treatments provided 78 to 83% control of smooth crabgrass in July, but control declined to <70% on September 8. Overall, sequential treatments of mesotrione plus prodiamine provided excellent (>90%) pre- and postemergence control of smooth crabgrass in tall fescue.

FLUCARBAZONE-SODIUM SELECTIVITY FOR PERENNIAL RYEGRASS CONTROL IN BERMUDAGRASS AND KENTUCKY BLUEGRASS. R. Singh*, S. Sidhu, T. Reed, and P. McCullough, University of Georgia, Griffin, GA (99)

ABSTRACT

Flucarbazone-sodium controls perennial ryegrass in bermudagrass and Kentucky bluegrass, but physiological effects attributed to selectivity have received limited investigation. Greenhouse and laboratory experiments were conducted to evaluate efficacy, absorption, translocation, and metabolism of flucarbazone-sodium in these grasses. Flucarbazone-sodium at 30 and 60 g ai ha^{-1} reduced shoot mass from the nontreated perennial ryegrass ≈22x and 3x more than bermudagrass and Kentucky bluegrass at 4 wk after treatment, respectively. In laboratory experiments, foliar and root absorption of ^{14}C-flucarbazone-sodium were similar among species. Bermudagrass distributed ≈25% more foliar absorbed ^{14}C to nontreated shoots, than Kentucky bluegrass and perennial ryegrass. From root applications, all grasses averaged 84% distribution to shoots. Bermudagrass and Kentucky bluegrass metabolized 100% and 74% of ^{14}C-flucarbazone-sodium at 1 d after treatment (DAT), while perennial ryegrass metabolism measured 44%, 58%, and 65% at 1, 3, and 7 DAT, respectively. Bermudagrass, Kentucky bluegrass, and perennial ryegrass recovered 4, 4, and 2 metabolites after 7 d, respectively. Results suggest differential metabolism of flucarbazone-sodium is attributed to selectivity for controlling perennial ryegrass in bermudagrass and Kentucky bluegrass.

A SURVEY OF GOLF COURSES PARTICIPATING IN THE METHIOZOLIN EXPERIMENTAL USE PERMIT PROGRAM. S.D. Askew*, K.M. Han, and S.J. Koo, Virginia Tech, Blacksburg, VA (100)

ABSTRACT

Methiozolin (MRC-01, PoaCure) has been successfully marketed in South Korea since 2010 and is nearing submission to the US Environmental Protection Agency(EPA) for possible registration. Methiozolin was first sprayed on US soil in Harrisonburg, VA on April 6, 2009 and is now under evaluation at over 30 US universities. In fall of 2011, S. J. Koo, CEO Moghu Research Center, devised a plan to collect names and addresses of golf course managers willing to participate in an experimental use permit (EUP) program to further improve methiozolin use patterns for golf putting greens, fairways, and tees, and investigate potential edaphic or environmental variables that might influence product efficacy or environmental impact prior to its expected registration. A team of university professors including James Baird, Univ. California Davis, Harold Walker, Auburn Univ., and Shawn Askew, Virginia Tech worked with S. J. Koo and Kyung Han, Product Development Manager, Moghu USA to create a mechanism by which superintendents across the US could be given the opportunity to participate in the EUP program. A survey was created to collect vital information required by EPA and information related to putting green management, edaphic variables, and amount and type of annual bluegrass infestation.

Superintendents from over 165 golf courses signed up for the EUP program in spring and early summer of 2012. The EUP application was submitted to EPA in fall 2012. The EUP will allow participants to purchase and use methiozolin under the trade name PoaCure on up to 6 acres per golf course. Based on survey responses, 56% of superintendents indicated they desired to use PoaCure on 4 to 6 acres and 91% indicated that greens and green surrounds were the desired treatment area. Superintendents provided input on 37 different parameters related to golf management on their courses. These data will eventually be compared to on-site performance of PoaCure at each of these courses in efforts to maximize product performance and improve product labeling for US, Japan, Canada, Australia, and other countries around the world. The survey information, however, already contains interesting information about how greens management practices may influence annual bluegrass infestations. A multivariate analysis of the principal components of average green size, green age, mowing height, mowing frequency, number of golfer rounds per year, maximum thatch depth, organic matter percentage, CEC, pH, nitrogen application per year, infiltration rate, latitude, and longitude was conducted. Principal components(PC) 1 and 2 were then regressed against annual bluegrass cover and the estimated percentage of annual bluegrass that was deemed to be the perennial biotype. Principal component 1 was positively correlated ($r^2 = 0.34$) to annual bluegrass cover and significant ($p<0.001$). The loading factors for PC1 were green age (0.7561), infiltration (-0.7047), CEC (0.4951), average green size (-0.3504), and organic matter (0.3402). A regression of green age by annual bluegrass cover indicates approximately 2.22% annual bluegrass is added for each year a green is in play in the first 15 years ($r^2 = 0.95$). Furthermore, as infiltration increases, annual bluegrass cover, organic matter, and CEC decrease. Few factors significantly influenced the percentage of annual bluegrass that was of the perennial biotype. A nonlinear relationship (r2 = 0.81) was noted between putting green mowing height and percentage perennial biotype indicating that higher heights of cut lead to less perennial biotype. Latitude and Longitude were not correlated to either annual bluegrass cover or percentage perennial biotype ($r2 < 0.002$) but did indicate that as one moves further north and east, mowing frequency increases from approximately 4 to nearly 12 times per week and number of rounds per year decrease. The data suggest that several other factors likely influence both percentage annual bluegrass cover and percentage perennial biotype. In addition, these data do not represent a random sampling of the golf courses in the US, but rather is biased toward ones that have problems with annual bluegrass. Thus, factors like geographic distribution might have less influence compared to a random population.

POSTEMERGENCE AND RESIDUAL CRABGRASS CONTROL WITH CAVALCADE PQ IN COOL-SEASON TURF. A.N. Smith*, S.D. Askew, and G.S. Hackman, Virginia Tech, Blacksburg, VA (101)

ABSTRACT

Quinclorac is a postemergence control option for crabgrass in cool-season turfgrass. Preemergence products containing quinclorac allow some flexibility in application by combining preemergence residual control with early postemergence crabgrass control from quinclorac. Products containing prodiamine and quinclorac were compared for efficacy against smooth crabgrass (*Digitaria ischaemum* (Schreb.) Schreb. ex Muhl.) and broadleaf weeds.

Two trials were initiated in fairway height perennial ryegrass (*Lolium perenne* L.) on June 12 and July 9 2013. On June 12, smooth crabgrass was 3-4 leaf stage. On July 9, smooth crabgrass was 1-2 tiller stage. Treatments were as follows: Cavalcade (prodiamine) at 0.73 kg ai ha^{-1}, Cavalcade PQ (prodiamine + quinclorac) at 1.68 kg ai ha^{-1}, quinclorac at 0.84 kg ai ha^{-1}, dithiopyr at 0.57 kg ai ha^{-1}, Cavalcade PQ at 1.34 kg ai ha^{-1}, and Cavalcade PQ at 1.34 kg ai ha^{-1} with methylated seed oil (MSO). A non-treated check was included for comparison. Crabgrass control was rated 1, 2, 4, 6, 8, 12, and 16 weeks after treatment (WAT).

In 3-4 leaf stage crabgrass, initial control was excellent for all products containing quinclorac. At 2 WAT, products containing quinclorac controlled crabgrass 91% or better. The best control was 100% from Cavalcade PQ + MSO and quinclorac alone. Dithiopyr controlled crabgrass 25%. At 4 WAT, products containing quinclorac controlled crabgrass 99% or better. Dithiopyr controlled crabgrass 94%. At 8 WAT, quinclorac alone and dithiopyr controlled crabgrass 78%. Cavalcade did not provide sufficient control. Cavalcade PQ at both rates and with MSO controlled crabgrass 94-98%. The same treatments controlled white clover 100% and common dandelion 88-98%. At 12 WAT, all Cavalcade PQ treatments controlled smooth crabgrass 90% or better. Quinclorac alone and dithiopyr controlled crabgrass 79 and 70%, respectively.

In 1-2 tiller crabgrass, dithiopyr and Cavalcade provided minimal crabgrass control; only treatments containing quinclorac controlled crabgrass. At 1 WAT, quinclorac alone and Cavalcade PQ controlled crabgrass 81-86%. At 2 WAT, crabgrass control ranged from 81-93%. Cavalcade PQ at 1.68 kg ai ha^{-1} controlled crabgrass 93%. At 4 WAT, crabgrass control ranged from 73-89%. Quinclorac alone controlled crabgrass 86%. Cavalcade PQ at 1.34 kg ai ha^{-1} controlled crabgrass 73%. Cavalcade PQ at 1.34 kg ai ha^{-1} + MSO and Cavalcade PQ at 1.68 kg ai ha^{-1} controlled crabgrass 88 and 89%, respectively. For 6, 8, and 12 WAT, treatments maintained similar control across dates. Cavalcade PQ at 1.68 kg ai ha^{-1} and at 1.34 kg ai ha^{-1} + MSO averaged 80-81% control. Quinclorac alone averaged 76-77%. Cavalcade at 1.34 kg ai ha^{-1} without MSO controlled crabgrass 60-70%. These results suggest that the inclusion of MSO to Cavalcade PQ at a lower rate can provide sufficient control of smooth crabgrass and can significantly increase crabgrass control when applied to more mature crabgrass.

THE EFFECTS OF TOPRAMEZONE ON CONTROLLING BERMUDAGRASS. C. Mansue* and J. Murphy, Rutgers University, New Brunswick, NJ (102)

ABSTRACT

Field experiment was conducted in 2012 to evaluate efficacy of topramezone for bermudagrass (*Cynodon dactylon* L.) control with combinations of triclopyr or quinclorac with comparison to fenoxaprop. All topramezone treatments contained methylated seed oil. Treatments were applied August 22, 2012 followed by two, 21-day interval applications. Treatments that contained topramezone alone showed good control (80%) on bermudagrass at 8 weeks after initial treatment (WAIT). Treatments that contained combinations of triclopyr or quinclorac showed between 55 to 70% control of bermudagrass at 8 WAIT. All treatments were followed up with a summer rating and showed that topramezone alone, plus triclopyr, or plus quinclorac had excellent control (99%) over bermudagrass compared to fenoxaprop (77%).

PREEMERGENT CONTROL OF SILVERY THREADMOSS PROPAGULES WITH HERBICIDES, FUNGICIDES, AND PLANT GROWTH REGULATORS. A. Post*, S.D. Askew, and D.S. McCall, Oklahoma State University, Stillwater, OK (103)

ABSTRACT

Silvery threadmoss (*Bryum argenteum* Hedw.) naturally reproduces through spore and bulbil production, both of which have potential to be controlled prior to establishment. Studies have not evaluated effects of turf protection products on moss protonema or gametophyte growth from spores or bulbils, and most moss is controlled postemergence on putting greens. Initial studies were performed to determine the optimal growth temperature for spores and bulbils in sterile culture. Protonemal tissue from spores grew optimally at 29.5°C and gametophytes from bulbils grew optimally 22.5°C. When evaluating the effects of turf protection products on moss development form spores or bulbils in axenic studies were conducted at a constant 24°C.

Bulbils and spores were collected from a putting green in Blacksburg VA in 2011 and 2012 plated onto 25% potato dextrose agar spiked with turf protection products including herbicides, fungicides and plant growth regulators. Three trials were conducted with three replications each. Plates were imaged at 0, 7, 14, and 21 days after initiation and analyzed in Sigma Scan Pro 5.2 for percent reduction in green pixels compared to the nontreated. Pixel counts over time were also summarized over dates to calculate reduction in area under the progress curve (AUPC) as a measure of silvery threadmoss control.

Carfentrazone is the industry standard for postemergence silvery threadmoss control on creeping bentgrass putting greens and it also reduced green cover of moss protonemata and gametophyte production from spores and bulbils in these studies. All combinations with carfentrazone reduced area under the progress curve (AUPC) for green cover of moss for both spores and bulbils by 80% or more by three weeks after treatment. Sulfentrazone, oxyfluorfen, oxadiazon, saflufenacil, flumioxazin, and pyraflufen-ethyl reduced AUPC of moss equivalent to carfentrazone for both propagule types. Two products containing fosetyl-Al, a phosphite, and mineral oil caused an increase in moss cover between 22 and 113% of the nontreated for spores; however, only methiozolin positively influenced AUPC (90.2%) compared to the nontreated for bulbils. Though silvery threadmoss is typically targeted postemergence on putting greens, there are labeled products available that can provide preemergence control, including the industry standard of carfentrazone. These data suggest that differences may occur between turf protection products in their ability to suppress moss establishment from spores or bulbils.

NEWSS YEAR-END FINANCIAL STATEMENT, 2012

NORTHEASTERN
WEED SCIENCE SOCIETY

NEWSS Fiscal Year Financial Statement for 2012	
November 1, 2011 to October 31, 2012	
New Revenue Received	***Amount***
Annual Meeting (Registrations, Dues only, Student Support, Late fees, BASF award, Industry Social)	$36,049
Interest Income (1st National Bank of Fredericksburg : Savings and CD's)	$955
Proceedings	$2,472
Standing Orders	$1,830
Sustaining Membership (Industry Dues)	$10,250
Weed Contest Support from Sustaining Members (Sponsorship)	$6,000
American Society for Horticulture Science (Share of annual meeting expense)	$824
Total Fiscal Year New Revenue	**$58,380**
Expenses	***Amount***
Administration (Stamps, Supplies, Set Up Fees For 2012 Meeting, New Computer, Printer, bank, adobe fees.)	$1,842
Annual Meeting (Facility Contract, Symposium speakers, Registration staff)	$20,099
Annual Meeting Awards (Student Competitions, Photo Contest, Fish Art Plaques, Logo Design)	$1,428
CAST Dues 2011	$1,500
Liability Insurance	$257
Proceedings	$3,900
Standing Orders	$901
Student Room Reimbursement	$2,490
Website	$575
Weed Contest Event Fees (Host Fee, Bus, Awards Banquet, 2012 Awards and 2011 Awards)	$12,104
WSSA Rep Travel to Summer Board Meeting	$730
CAST Rep Travel to CAST Conference	$1,019
Total Fiscal Year New Expenses	**$46,845**
Profit/Loss	
+$11,536	
Assets Values on October 31, 2012	
Cash	$0
1st National Bank of Fredericksburg Checking Account (October 12,2012)	$9,525.84
1st National Bank of Fredericksburg Savings Account (October 12,2012)	$31,765.71
2012 CD Weed Contest Support in Reserve (October 12,2012)	$24,733.67
2016 CD Annual Meeting in Reserve (October 12,2012)	$10,090.80
Total Net Worth on October 31, 2012	
	$76,116

Signature of Preparer: ______________________ Date ______________

Accounts were reviewed by the undersigned and are in order.

Jim Steffel ______ Date 02/06/2013 Todd Mervosh ______ 2/6/13

Art Gover ______ Date 2/6/2013 Erin Hitchner ______ Date 2/6/13

NEWSS YEAR END REPORT 2012

Submitted by the Executive Committee
For the 67th Annual Meeting
February 6, 2013
Hilton Baltimore
Baltimore, MD

President's Report – Antonio DiTommaso

The 66th Annual Meeting held in Philadelphia, PA at the Hyatt Regency at Penn's Landing, January 3-6, 2012 was very successful. Attendance was higher than the past two years and within the average of the previous five years. Our Past President, Mark VanGessel, did a masterful job in securing a different hotel for our meeting once we learned that renovations to our scheduled hotel would not be completed in time just one week and a half prior to the start of our annual meeting. Again, this year we met jointly with the Northeastern section of the American Society of Horticultural Science (NE-ASHS) who actively participated in the General Sessions and workshops. The theme of the meeting was "*Weed Science Funding: New Avenues for a Changing Landscape*". We had two excellent workshops, one titled "Moving Beyond Rotating Modes of Action: Advancing the Concept and Application of Integrated Weed Management" and the second symposium –was a mini-symposium focused on our current understanding of the distribution, ecology, and management of the invasive swallow-wort vines.

At the 2012 annual meeting, the NEWSS membership also approved the splitting of the Treasurer and Secretary positions, the latter of which will be renamed the "Membership Chair" position. The membership also voted to adopt a new logo for our society (see front page of this report).

Award recipients at the 2012 annual meeting were:

Ronald Ritter, Award of Merit
Timothy Dutt, Fellow Award
Sherrilynn Novak, Service Recognition Award
Robert D. Sweet Outstanding Graduate Student Award:
Matthew A. Cutulle (Ph.D.)
Jose Javier Vargas (M.S.)

Graduate Student Presentation Awards went to: K. Averill (First place) and D. Tekiela (Second place). The Graduate Poster Awards went to: B. Crockett (First place) and K. Venner (Second place). Photo Contest winners were J. Vargas (First place), M. Bravo (Second place), and J. D'Appollonio (Third place).

New members to the Executive Committee joining in 2012 were Greg Armel as Vice President, Javier Vargas, Public Relations; and Erin Hitchner as our incoming Treasurer who worked closely with Melissa Bravo during the year, prior to assuming full responsibility for the position. During 2012, we also welcomed Keith Burnell in the newly created position of Membership Chair and Bryan Dillehay as Research and Education Coordinator.

NEWSS continued to be on a sound financial footing thanks to the hard work of Melissa Bravo and Mark VanGessel and the entire membership. Once again, John O'Barr did an excellent job in soliciting support from our sustaining members. Hotel rates and registration rates for the 2012 meeting were at the same levels as in 2008.

The 2013 NEWSS annual meeting will be held as a joint meeting with the Weed Science Society of America (WSSA) in Baltimore, MD at the Hilton Baltimore Hotel from February 4-7, 2013. Our President-Elect, Dwight Lingenfelter, has worked closely with WSSA program Chair, Jim Kells, to develop a strong and varied program.

The 2012 NEWSS Collegiate Weed Contest was held on July 25 at Rutgers' University Clifford E. and Melda C, Snyder Research and Extension Farm facility in Pittstown, NJ. John Grande and his entire staff and volunteers did a masterful job in assuring a successful contest. More than 65 graduate and undergraduate students from 7 institutions from as far north as Canada (Guelph), and south, as North Carolina competed in this year's contest. Institutions represented at the contest included Cornell Univ., North Carolina State Univ., Ohio State Univ., Penn State Univ., Univ. of Kentucky, Virginia Tech, and Univ. of Guelph.

The Executive Committee met briefly in January at the annual meeting and had three phone conferences, one in January as a summary of the 2012 annual meeting, one in March and again in November. The November phone conference meeting replaced the customary October meeting at the upcoming meeting hotel location since the WSSA was responsible for all hotel related tasks for the joint 2013 NEWSS/WSSA meeting. The Executive Committee also met in July at the NEWSS Collegiate Weed Contest location in New Jersey. Committee Lists were updated and posted to the NEWSS website. Changes were made to the Manual of Operating Procedures. No resolutions were brought by the Resolutions Committee in 2012.

The Executive Committee unanimously approved pursuing the establishment of a NEWSS Endowment that had been proposed by the NEWSS Past Presidents including Renee Keese, Jeff Derr and Hilary Sandler. This endowment would support future educational endeavors of the society including graduate student support. The President created an ad-hoc committee to determine what would be needed to establish this foundation including objectives, goals and oversight of the foundation. Greg Armel agreed to Chair this ad-hoc committee which included Melissa Bravo, Jeff Derr, Erin Hitchner, Renee Keese, John O'Barr, and Hilary Sandler. The establishment of a NEWSS Endowment and its associated committee was put to an electronic vote of the membership in December 2012 and by an overwhelming majority of 98% the NEWSS membership elected to establish a NEWSS Endowment. The ad-hoc committee will update the membership on their progress toward implementing their plan for the Endowment at the NEWSS Annual Business Meeting on February 6, 2013.

I have truly enjoyed the opportunity to serve as your President during this past year. I have learned much from all of you and have been impressed with the level of passion and commitment that you all have for our society. I want to specifically recognize Darren Lycan who will be rotating off the committee as Editor and will be replaced by Art Gover and Melissa Bravo who will rotate off the committee as the Treasurer/Secretary after her excellent years of service. I would also like to thank John Willis for his service as Research and Education coordinator and Javier Vargas as Public Relations representative. Finally, I would like to thank Mark VanGessel for his service and the wisdom he provided as Past President.

President Elect's Report – Dwight Lingenfelter

2012 Misc. Activities

- I worked with the other Executive team members to populate the 2012 Committee List
- Assisted T. DiTommaso and M. Bravo with developing a budget
- Generally, assisted the President, when necessary, with various activities throughout the year
- I have been serving as the NEWSS representative on the WSSA Public Awareness Committee for the past several years; involvement requires routine conference calls, technical input and editing of monthly press articles

2013 Co-Program Chair for Joint Meeting

- I routinely communicated with Jim Kells, WSSA Pres-elect, about logistics for 2013 joint meeting (ie, overall program, special needs of NEWSS, student paper and poster contest, committee meeting rooms, etc.)
- Also, I was in communication with Joyce Lancaster, WSSA Executive Secretary and Kate Counter, WSSA Meeting Coordinator about various meeting issues
- 24 student oral presentations will comprise the NEWSS speaking contest during the annual meeting; 13 students will be in the NEWSS poster contest
    - Details about this were conveyed to the students in newsletters, emails and the like, so to provide updates on logistics
- Otherwise all of our sessions, committee and business meetings, and Past Presidents breakfast will fit into the standard WSSA program
- Overall, it has been a pleasure working with WSSA on this event

2014 Annual Meeting Venue

- I made two hotel site visits, both in Philadelphia: Sheraton Society Hill and Hyatt Regency
- We have recently held our annual meeting at both locations
- After consideration and discussion with the site selection committee, I am in final negations with Sheraton Society Hill, and plan to have a contract signed before the joint meeting in Baltimore
- Sheraton Society Hill had lower room and parking rates, is located closer to restaurants and other attractions, and in general seems more willing to work with

our needs for future meetings; however, Hyatt was competitive and should be considered if we decide to return to the Philadelphia area

- NE-ASHS wants to join us for a joint meeting again in 2014; I am working with them as we prepare for the meeting

Secretary's Report – Melissa Bravo

Banking information: Signatories on the bank accounts are primary* -Melissa Bravo, secondary-Antonio DiTommaso, third-Dwight Lingenfelter and fourth Erin Hitchner. Due to increasing bank security precautions it will be necessary for outgoing EC board members to update the signatory cards immediately after the annual business meeting. The NEWSS also has a limited daily total transaction use ($1,000.00) mac card issued to the primary holder of the account (the Treasurer). Please note that this bank charges a $15.00 foreign (Canadian and Euro) check processing fee unless the check is drawn on a United States Bank or received as a money order in American dollars and a $5.00 fee for cashier's checks and $12.00 fee to void or cancel a check. *As the EC treasurer position rotates off the bank signatory cards, accounts, post office box will be changed to Erin Hitchner, effective February 15, 2013. Change of address forms, letterhead and updating EC OFFICER information with insurance policies, tax exempt forms and DUNS number will need to be made.

Insurance: The Society holds a Dishonesty Bond Insurance Policy with Fingar Insurance of New York for the E.C Board. This policy is current and remains in effect through February 16, 2013. The policy renewal has been sent to Fingar Insurance. At this time the Society does not hold a Liability Policy. Instead, single event (Weed Contest, NIVM short course) liability policies were purchased when requested by the Event Facility. Therefore, any event that requires NEWSS be bonded for liability must be sent to the Secretary 6 months prior to the event in order to acquire a Liability Policy (*e.g. as done for NIVM in 2008 and 2009*).

Standing Orders: Current year and archived *Proceedings* continue to be received by Irene T. Bradley and shipped from Cornell University. Past proceedings are also scanned and archived on the web.

Web site: Updates and maintenance continues to be provided by Rob Dickerson at Penn State University. EC members worked with Rob to post the updated EC Contacts, annual meeting registration links and other society information.

New Logo: The winning logo contest image has been incorporated into the EC letterhead and awards. Updates to the website, banner and other promotional materials will continue through 2013.

Office Equipment: A laptop, colored printer/copier and zip drives, and virus protection were purchased in 2012 and will be kept and maintained by the current Treasurer. Items will be brought to the annual meeting for registration and event office support needs.

Weed Contest: Fees for this year's contest were paid to Rutgers University for hosting the event, First Student Bus Garage for transportation for students, the event banquet hall fee and event awards. The total event costs came to $11,154.26. Monies received by NEWSS for the Contest were $6,500 from Sustaining Platinum and Gold members for year 2012; $2,000 from sustaining members that carried over from year 2011; and

$1,000 from an un-deposited sustaining check from year 2007 that was reissued to us this summer. The remaining balance ($1,654.26) was provided by the EC on behalf of the membership at large. The fees were paid from checking and it was not necessary to cash out the Weed Contest CD that matured in August of 2012 renewed (3/23/2014) but at a very low interest rate (.30000% AIR). The EC voted in July to increase this reserve to a total of $15,000.00. This reserve is meant to cover Weed Contest expenses over and above the annual contributions received from our Platinum and Gold level industry sustaining members who sponsor the event.

Joint Annual meeting: Registration fees and forms were approved by the EC and posted to the website in November. WSSA was approved to accept and process our annual meeting registrations and dues through their website and the WSSA will send a check to the NEWSS after the event (less expenses). Purchase of proceedings will be available at the annual meeting NEWSS information desk and on line at ACTEVA from January 18th through February 15th. Proceedings will be shipped in March this year.

Hotel Tax Exempt Form: The NEWSS currently has a valid tax exempt certificate forPA (expires 2-2015); MA (expires 10-2018); MD (expires 9-30-2017).

Membership update: As of November 2012, there are 300 **members-at-large** in the Updated Directory (2010-2012 Meeting Attendees) and additional contacts in the NEWSS Listserve who do not consider themselves regular members of the society. Members-at-large who have not been active in the society in the past 3 years are removed from the printed directory if not registered at the next annual meeting, or if they have not paid member dues only ($50.00) for the current fiscal year, or have since passed away.

NEWSS Annual Meeting	2005	2006	2007	2008	2009	2010	2011	2012*
Distinguished/Fellow	11	4	12	11	11	8	11	2
Dues only	Data not available				6	12	11	9
Invited Speakers		3	3	5	3	7	7	5
Regular Members	200	139	148	119	132	91	104	110
Retired or Retired and Distinguished/Fellow	7	5	4	1	3	3	3	6
Student members	50	23	19	21	32	19	21	29
Sustaining Companies	.	16	22	22	13	16	21	17
Symposiums and Workshops	172	22	39	14	76	0	18	20
TOTAL	314	212	247	193	276	156	196	197**
Proceedings Distributed	118	> 116	79	65	75	50	78	117
Standing Order Sales	50	50	50	50	50	50	55	46

*51% of our 2012 meeting registrations were processed through ACTEVA and paid additional for ACTEVA fees.

** 66% of our at large members (Membership Directory 2010-2012) attended or sponsored the annual meeting in 2012; conversely, 34% of our at large membership did not pay dues or registration fee in 2012.

Vice-President's Report – Greg Armel

By an overwhelming majority the NEWSS membership (97.6%) voted to support the initiation of an NEWSS endowment fund and corresponding oversight committee. Seed money for this endeavor will be provided from any small remaining funds (after spending on necessary appropriations) generated from our joint meeting with the WSSA. The rest of the start-up funds for this endowment will need to come from generous donations of our membership. This endowment will take several years to fully establish however it is our goal for our endowment to provide some small but visible support following our first year of establishment. Our endowment committee including myself, Renee Keese, Jeff Derr, John O'Barr, Melissa Bravo, Hilary Sandler, and Erin Hitchner will spend the next year petitioning the membership to financially support this new endeavor. Following our year of service in its establishment, the endowment committee will be annually chaired by the current Vice-President of the society with the remaining committee members being the three most recent Past Presidents, the Treasurer, and one member at large appointed by the current President. The oversight committee's responsibility (with advice and consent by the board) is to raise money, invest donations, and advocate for expenditures of surplus funds in the endowment for support of student programs. It shall be noted that the endowment will be set up as a non-profit entity that will follow all legal guidelines established for such organizations and will remain a function of the NEWSS and will not act outside of the oversight of the NEWSS board. It is our hope that this new established endowment will set the foundation for future support for student programs for our society.

In addition as we approach our 2014 meeting, we are always in need of great ideas and insights for workshops and symposia. If you have any topics of interest, or just some general ideas about the meeting please let me know (gregory.armel@basf.com) so we can organize a productive educational event. Also for our 2015 meeting, I would like to take the NEWSS to a new venue either in or outside of our current geography. Be on the lookout for a survey where I will ask you to vote for your favorite venues for the meeting. Your vote matters greatly so stay tuned.

Past-President' Report – Mark VanGessel

The Secretary/Treasurer position was split based on vote at the 2012 NEWSS business meeting. The ByLaws were amended to reflect this change. Several changes were made in the MOPs around splitting the Secretary/Treasurers responsibilities to the Treasurer and Membership Committee.

The minutes from the 2011 Executive Committee meetings, annual business meeting, financial statements, newsletters, and other relevant publications were sent to the archives (Dan Kunkel) at the IR-4 Project Headquarters archive room in NJ. Information on the new NEWSS logo and the split of the Secretary/Treasurer position was also included.

EC voted to rename the Service Recognition award, the M. Garry Schnappinger-Service Recognition Award. The Awards Committee consisted of Mark VanGessel as chair, Hilary Sandler, Dave Yarborough, Jerry Baron, and Renee Keese. Nominations were received for Fellow, Award of Merit, Outstanding Researcher, Outstanding Educator,

Service Recognition, and Graduate Student (both MS and PhD). Plaques were ordered for all award winners with the updated NEWSS logo and font. The Awards program is being held during the opening session of the NEWSS/WSSA joint meeting. WSSA is publishing the Awards Brochure with bio's for all award winners.

The Past-Presidents' Breakfast is scheduled for Wednesday February 6, 2013 at the Joint meeting.

Editor's Report – Darren Lycan

Titles and abstracts were submitted via the WSSA submission website for the joint 2013 annual meeting of the NEWSS and the WSSA. Program booklets contained information for both societies for the fully integrated joint meeting and were prepared by the WSSA. Only abstracts that were submitted by NEWSS members and other NEWSS information is included in the 2013 NEWSS Proceedings. There were 45 NEWSS titles submitted for Posters and 93 NEWSS titles submitted for Oral presentations. Due to the delayed submission deadlines for the 2013 joint meeting, printed Proceedings will not be available at the meeting in February, but will be mailed to those who order a copy after the meeting. The cover of the 2013 Proceedings will feature the winning photo from the 2012 NEWSS photo contest: henbit (*Lamium amplexicaule*) taken by Javier Vargas of the University of Tennessee. At the time of this report, the final version of the Proceedings has not been submitted to Omnipress for printing. Likewise, ordering is still active so a final count and cost of printing has not been determined. This specific information will be included in a subsequent Editor's report.

Sustaining Membership Report – John O'Barr

A request for Sustaining Membership was sent to all current sustaining members and prospective new members in fall 2012. We have received commitments from 19 organizations for 2013. This is an increase of one organization from 2012 as well as an increase in sustaining revenue of $250 over year.

The 2013 Sustaining Membership breakdown is as follows: 2-Platinum, 6-Gold, 5-Silver, and 6-Bronze confirmed for a total commitment of $15,000.

2013 NEWSS Sustaining Members

Organization	**2013 Level**
BASF Corporation	Platinum
Syngenta	Platinum
Bayer Crop Sciences	Gold
Dow AgroSciences, LLC	Gold
United Phosphorus, Inc.	Gold
FMC Corporation	Gold
Monsanto	Gold
Valent	Gold
PBI Gordon	Silver

OHP	Silver
BAAR Scientific	Silver
For-Sure Weed Control, Inc.	Silver
DuPont	Silver
ACDS Research	Bronze
Gylling Data Management	Bronze
TeeJet® Technologies	Bronze
Crop Management Strategies (CMS)	Bronze
WEEDS, Inc.	Bronze
LABServices	Bronze

2013 Sustaining Members will be recognized at the 2013 annual meeting and at the 2013 collegiate weed contest. No organization has yet requested a commercial display table for the 2013 annual meeting in Baltimore.

I would like to personally thank each sustaining member for their generous support of the Society.

WSSA Representative Report – Jacob Barney

- The 52nd annual meeting was held in Kona, Hawaii.
- NEWSS is joint hosting the 2013 meeting with WSSA in Baltimore, MD – Jim Kells is organizer, and Scott Glenn is serving as social arrangements chair.
- WSSA is looking to replace EPA Subject Matter Expert Jill Schroeder who is stepping down after a wonderfully successful tenure in this new position. Contact Rod Lym for information or interest (Rod.Lym@ndsu.edu)
- Website to get major redesign. This was decided and agreed upon this year, but no progress has been made on implementation yet. WSSA recently hired a website content manager (Muthu Bagavathiannan, postdoc at Univ Arkansas) who will lead an effort to keep the web content up to date and well organized. Hopefully the WSSA website will not only be better organized, but will be a source for information for weed scientists.
- National Invasive Species Awareness Week was Feb 26-Mar 2, 2012 with over 800 participants. Several WSSA members were involved in this important outreach and education event. Rod Lym gave the opening presentation on the successful management program "Team Leafy Spurge" (nisaw.org)
- WSSA is interested in supporting an effort to develop a CAP (Coordinated Agricultural Project) within USDA for weed science.
- WSSA is looking to reduce the number of Members at Large serving on the Board of Directors to reflect the lower membership number.
- A special committee was formed to investigate the development of an "alternative" form factor for the Herbicide Handbook. Perhaps a tablet or e-version are on the table as potential new products to keep with technology advancements.
- Future WSSA Meeting Sites:
  2014 Vancouver, British Columbia as joint meeting with Canadian Weed Science Society
  2015 Lexington, Kentucky
  2016 San Juan, Puerto Rico
  2017 Tucson, Arizona

Graduate Student Representative Report – Adam Smith

The graduate student events for the WSSA/ NEWSS joint meeting will be sponsored by Monsanto and will include a graduate student luncheon and a professional development workshop. Both are being held on Wednesday. A new NEWSS student representative will be voted into office at the luncheon; there are three candidates. Each candidate will be allowed a brief statement and the NEWSS students will have the opportunity to vote.

The NEWSS paper and poster contests will be held on Tuesday. The paper contest has morning and afternoon sessions. Awards will be given for each session. All talks must be uploaded at the registration desk by 4pm Monday. Exceptions will be made for those that are traveling at the time. There is a point deduction for those who do not meet this deadline. Winners for these contests will be announced at the NEWSS business meeting. Judging criteria for both paper and poster contests can be found on the NEWSS website. Poster size was changed for the joint meeting; posters *cannot* exceed 48x48.

NEWSS CAST Representative Report – Randall Prostak

News from CAST

CAST continues to search for funding in difficult economic times.

Dr. John M. Bonner, Executive Vice President/CEO, has announced his retirement in January 2013. A search for the next EVP/CEO is being lead by Phil Stahlman and CAST Board of Directors

On November 15, 2012, the Board of Directors of the Council for Agricultural Science and Technology has appointed Linda Chimenti, current Chief Operating Officer (COO) of CAST, as the next Executive Vice President of the 40-year-old organization. Dr. John Bonner, current Executive Vice President/Chief Executive Officer (EVP/CEO), will leave that position at the end of January 2013 to pursue other interests. Ms. Chimenti has years of experience as a CAST administrator, and she is committed to the organization's mission. As she says, "It is an honor to have the opportunity to take on a new role in CAST, an organization that I have served for twelve years with growing responsibilities." Ms. Chimenti also considers the move a compliment to the entire CAST staff. "The Board of Directors has expressed their confidence in our ability to maintain CAST's output of high quality, credible publications and programs." After working at Iowa State University's Center for Agricultural and Rural Development, Ms. Chimenti joined the CAST organization in January 2001 as Managing Scientific Editor and has led the editorial department continuously since that time. During this period, she has been responsible for the production of approximately 140 CAST publications, including Task Force Reports, Issue Papers, CAST Commentaries, Annual Reports, and organizational histories. In 2008, Ms. Chimenti was promoted to Director of Council Operations and added staff supervision and management duties to her responsibilities. She assisted the Board of Directors and Board of Representatives with a major restructuring of the organization's governance in 2009. Since 2010, Ms. Chimenti has served as COO, working closely with Dr. Bonner on initiatives to expand CAST's impact and broaden its audience. According to Dr. Phil Stahlman, CAST President, "We are very pleased that this transition of leadership will take place within the CAST organization now that Dr. Bonner is stepping down to pursue other interests. We are deeply grateful for all he has done for CAST, and we wish him all the best. We welcome Ms. Chimenti to her new role and look forward to working with her in this capacity."

On September 26, 2012, Dr. Carl Winter Presented with the Borlaug CAST Communication Award. If consumers and policymakers are not getting the message about agriculture, it's not Carl Winter's fault. The respected professor from the University of California--Davis is the author of two books and more than 100 publications. He serves on committees, teaches communication courses to graduate students, and testifies before the U.S. Congress on pesticide/food safety issues. But he is maybe best known for his version of America's Got Talent--Ag Style.

More information of the activities of the past year and be reviewed at:

http://www.cast-science.org/about/summary_of_activities/

News from the NEWSS CAST representative

The 2012 CAST board meeting was held November 12-15 in Kansas City, MO

Western Society of Weed Science's Phillip Stahlman, professor of weed science at the Kansas State University, became CAST president for 2012-2013.

Randy Prostak finished his term as chair of Plant Agriculture and Environmental Issues Working Group in October 2012. At the request of CASTleadership, he will chair a new committee that has been created to mentor individuals new to the CAST board. Randy will also serve on the 2013 CAST nominating committee.

Still a possibility that the 2013 CAST board meeting will come to the northeast.

2012 CAST Publications

Water and Land Issues Associated with Animal Agriculture: A U.S. Perspective
Issue Papers - IP50 - August 2012

This paper addresses specific water and land concerns related to animal agriculture. The authors consider issues of water use and water quality associated with the livestock sector and the related environmental and economic impacts. The paper also includes a discussion of livestock land use, land degradation, land application of manure and manure-based composts, and deforestation related to farm policies. This Issue Paper presents scientific data specific to North America (primarily the United States) and explains how and why these data compare with other parts of the world. Additionally, it discusses how to reduce the environmental impact of livestock agriculture while preserving resources and quality of life. Chair: Kelly D. Zering, North Carolina State University

The Direct Relationship between Animal Health and Food Safety Outcomes
CAST Commentaries - QTA2012-1 - May 2012

The health of the animals within the food animal production system impacts many aspects of the system far removed from the animals themselves. To promote high productivity in animal agriculture, researchers need to examine nutrition, management systems, and animal care practices including the use of antibiotics and vaccines. This Commentary looks at the pressures to change livestock rearing methods, evidence to support the direct public health impact on human illness days, and good safety and inspection service regulations. The authors use indirect evidence, diagrams, and graphs to deliver their findings about the ways that healthy animals result in safer food. Chair: Dr. H. Scott Hurd, Iowa State University

Assessing the Health of Streams in Agricultural Landscapes: The Impacts of Land Management Change on Water Quality Special Publications - SP31 - March 2012

The Clean Water Act of 1987 states that the elimination of pollutant discharge into navigable water is a national goal. Agriculture has been identified repeatedly as a significant contributor of nonpoint source pollution to surface waters. Despite conservation efforts, water quality problems still exist; all agricultural states struggle with balancing productive landscapes and water quality. Legislation, potential regulations, or allocations of millions of dollars to change agricultural practices seem warranted only if we know stream water is favorably impacted by modified agricultural practices, and in which streams the greatest potential impact might be observed. Project Manager/Lead Author: Rick Cruse, Iowa Water Center, Ames, Iowa.

Herbicide-resistant Weeds Threaten Soil Conservation Gains: Finding a Balance for Soil and Farm Sustainability Issue Papers - IP49 - February 2012, additional online video

Herbicides were developed during the twentieth century to be used with conventional tillage for weed control. Conservation (or minimum) tillage subsequently evolved, which enabled less soil damage when used with herbicides. Selection pressure, however, has resulted in weed species that have made adaptations for survival in conjunction with tillage. The U.S. government has put several federal policies and programs in place that help determine the selection and implementation of crops and conservation programs in relation to herbicides and tillage. This Issue Paper examines the impact of certain weed management practices on soil conservation objectives and addresses ways to mitigate negative effects. Chair: David R. Shaw, Office of Research and Economic Development, Mississippi State University, Mississippi State. IP49, February 2012, 16 pp.

Energy Issues Affecting Corn/Soybean Systems: Challenges for Sustainable Production Issue Papers - IP48 - January 2012

Energy Issues Affecting Corn/Soybean Systems: Challenges for Sustainable Production speaks to energy issues within the corn/soybean production system as a model for understanding the complexity of addressing global energy challenges. Research and development is needed to find ways to lower adoption barriers for energy-conserving practices and develop management systems that allow agricultural production to meet multiple demands. The challenges include (1) a growth in biofuel production directly from corn/soybeans coupled with a simultaneous growth in oil and agricultural commodity prices, and (2) the ability of the motor fuel infrastructure to handle an increased volume of ethanol, biodiesel, and advanced biofuels. The authors propose a landscape vision for sustainable corn/soybean systems that is feasible and could be done efficiently and economically if there is a desire and public willingness to do so. It would, among other things, provide sustainable grain and biomass feedstock supplies for the bioenergy industry, protect water quality, lessen producer/environmental risk, and promote biodiversity. Chair: Douglas L. Karlen, USDA—ARS, National Laboratory for Agriculture and the Environment, Ames, Iowa. IP48, January 2012, 16 pp.

Forthcoming Publications

Agricultural Innovation to Sustainably Feed the World by 2050 (Issue Paper)

Animal Feed vs. Human Food: Challenges and Opportunities in Sustaining Animal Agriculture Toward 2050 (Issue Paper)

The Contributions of Pesticides to Meeting the Global Need for Food Production and Public Health by 2050 (Issue Paper)

Food, Fuel, and Plant Nutrient Use in the Future (Issue Paper)

Impact of the Precautionary Principle in Feeding Current and Future Generations (Issue Paper) (Chair) Gary Marchant, Arizona State University

Integrated Nutrient Cycling: Key to Long-term Food Production (Issue Paper)

Omega-3 Fatty Acids: Health Benefits and Dietary Recommendations (Special Publication)

Research and Education Coordinator – Bryan Dillehay

Requests were sent to the following for pesticide recertification to be offered at the 2013 annual meeting. Current approval status is below, with 'n/a' indicating credits were not requested.

State	Approval	State	Approval	State	Approval
CT	Yes	NC	Yes	RI	Yes
DE	Yes	NH	n/a	VA	Yes
MA	Yes	NJ	Yes	VT	Yes
MD	Yes	NY	n/a	WV	Yes
ME	Yes	PA	Yes	CCA	Yes

With the upcoming joint meeting between NEWSS and WSSA, the program is significantly larger than years past. After a discussion with Dwight Lingenfelter and Bill Curran, I applied for recertification credits for selected sessions only and not the entire program.

Credits will be offered in the following sessions:
Herbicide Application in the 21st Century
NEWSS Graduate Student Paper Contest
Pasture, Rangeland, Forest, and Rights of Way
Integrated Pest Management
Herbicide Resistance Management – The Path Forward
Formulation, Adjuvant and Application Technology
Agronomic Crops
Turf and Ornamentals
Environmental Assessment of Herbicide Resistance
Horticultural Crops

I have not applied for recertification credits in New York and New Hampshire. The NH application process is arduous, the size of the upcoming meeting is significantly larger than years past, and lastly, no credits were given during the 2012 annual meeting. With regard to NY, Russ Hahn strongly recommended to not apply for NY credits this year due to the difficult application process and time constraints. For example: NY requires each session speaker to be approved before a session can be approved, which requires a resume or CV to be submitted. Once a speaker is approved, NY DEC requires session applications 45 days prior to the event. As such, recertification was not sought.

Director of Science Policy Report – Lee Van Wychen

Summary of Activities

1. Generated support for USDA research funding through competitive grants (AFRI) and formula funds (Hatch, Smith-Lever) as well as integrated programs (Regional IPM): met with House and Senate committee staff and wrote coalition letters on behalf of the National and Regional Weed Science Societies. FY 2013 budget is in limbo and a continuing resolution based on FY 2012 may fund the entire FY 2013. The current funding for FY 2013 is funded through March of 2013.
    a. President's Council of Advisors for Science and Technology (PCAST) recommended that the U.S. increase its investment in agricultural research by a total of $700 million per year:
        i. $180 million for new graduate and post-doctoral fellowships;
        ii. $235 million for new competitively funded research at USDA AFRI;
        iii. $130 million for basic research at NSF; and
        iv. $150 million for new public-private institutes.
    b. Top challenges they listed for 21st century agriculture:
        i. Managing new pests, pathogens, and invasive plants.
        ii. Increasing the efficiency of water use.
        iii. Reducing the environmental footprint of agriculture.
        iv. Growing food in a changing climate.
        v. Managing the production of bioenergy.
        vi. Producing safe and nutritious food.
        vii. Assisting with global food security and maintaining abundant yields.
2. Worked to prevent EPA from finalizing a rule which would allow fuel made from two known noxious weeds, *Arundo donax* (giant reed) and *Pennisetum purpureum* (napier grass or elephant grass), to count toward federally-mandated renewable fuels targets. We need more research on *Arundo* and *Pennisetum* before incentivizing their production. If OMB signs off on EPA releasing the rule, we requested to see a complete assessment of the costs and benefits, as outlined in Executive Order 13112. Additionally, if EPA approves Arundo donax and similarly high risk feedstocks, we believe that the rule must include – at the very minimum— guidelines on stringent best management practices to reduce the risk of escape.
3. Worked with interested coalitions to get support in both the House and Senate for the NPDES legislative fix bill, H.R. 872 and S. 3605. However, legislation remains blocked by Sen. Boxer (CA). Submitted letters of support on behalf of the National and Regional Weed Science Societies. New legislation will have to be reintroduced in the 113th Congress.
4. Generated support for the Pesticide Safety Education Program (PSEP) through a technical paper, and a joint press release with entomology and plant pathology. Currently

participating in a national stakeholder group to find more permanent funding mechanisms at the federal and state level for PSEP.

5. Generated support for Army Corps of Engineers Aquatic Plant Control Research Program (APCRP) and sent letters to House and Senate Energy and Water appropriations committee members. FY 2013 funding in limbo again for APCRP. Senate has proposed $4 million, but House and President proposed $0. Senate support from Schumer (NY), Leahy (VT) and Cochran (MS) will be key again. I will need to send another letter to House and Senate appropriators before the end of March.
6. Coordinated and co-organized support for National Invasive Species Awareness Week activities which occurred Feb. 26 to March 3, 2012. Planning for NISAW 2013 is in the final stages. WSSA is hosting a presentation on Capitol Hill on Mar. 4, 2013 regarding invasive plants and biofuels. We are also working with other NGO's to introduce a Congressional resolution in the House and Senate supporting NISAW.
7. Supported Public Awareness Committee activities and responded to press inquiries. Recent headlines include six news releases on pesticide stewardship as well as the following:
    a. Three Leading Scientific Societies take an Objective Look at the Issues Associated with "Least Toxic Pesticides" Applied as a "Last Resort".
    b. Annual Meetings of Weed Science Societies to Highlight Latest Developments in Research and Management of Weeds and Invasive Plants.
    c. Decades Old Weed Seeds Trigger New Outbreak of Devastating Plant Parasite.
8. Continued educating Federal agency and NGO stakeholders on herbicide resistance management. Watched and listened for any legislation that would attempt to regulate herbicide resistance or restrict the interstate movement of herbicides due to resistance issues.
9. Helped plan and coordinate 2012 EPA Herbicide Resistance Education Tour on the Delmarva Peninsula in August. The tour report is on the WSSA website.
10. Participated in Farm Bill stakeholder meetings and advocated for the following provisions where differences existed between the House and Senate:
    a. Support the Senate provision that establishes the Foundation for Food and Agricultural Research (FFAR), a new nonprofit corporation designed to supplement USDA's basic and applied research activities, and provides total mandatory funding of up to $100 million of matching funds. The foundation will solicit and accept private donations to award grants or enter into agreements for collaborative public/private partnerships with scientists at USDA and in academia, non-profits, and the private sector.
    b. Strongly oppose the House provision that require matching funds for applied research and extension that is commodity or state specific.
    c. Support the Pesticide Registration Improvement Act (PRIA III) in the House bill. PRIA was set to expire on Oct. 1, 2012, but separate legislation was introduced and passed in Sept 2012.
    d. Support reauthorization of the Specialty Crops Competitiveness Act through FY2017 in the Senate bill. Increases mandatory funding to $70 million annually (FY2013 - FY2017), which would also raise the minimum grant amount received by each state/territory.
    e. WSSA supports conservation compliance, but only if exemptions are granted for herbicide resistance management.

The DSP and Science Policy Committee were active with many other issues on a continuous basis in 2012. Notable issues that arose during the past year that required support through letters, phone calls, and meetings included:

a) asked USDA NIFA to rethink their "Crop Protection Program" funding line item and explore options on how to maintain equal funding for the six programs involved, especially IR-4, the Regional IPM Centers, and Extension IPM.
b) Supported a letter to Congress asking them to avoid legislation that would place severe restrictions on government employees' abilities to attend meetings and conferences.
c) asked EPA OPP to support the use of MSMA in turf weed control

2013 Plan for Committee Activities

- Discuss weed science priorities with NIFA Director, Sonny Ramaswamy and generate support for USDA to fund a $10 million CAP grant for weed resistance management.
- Investigate the "fine print" of provisions that get incorporated in the 2013 Farm Bill.
- Investigate FY 2014 federal budget proposal for programs affecting weeds and invasive plants
- Support funding for the Aquatic Plant Control Research Program
- Monitor any new Renewable Fuel Standard proposals that try to add invasive plants as biofuels.
- Continue support for an NPDES Fix bill in the 113th Congress.
- Advertise and hire a new EPA Subject Matter Expert (SME) to replace Jill Schroeder.
- Explore the possibility of establishing a similar SME position within USDA
- Engage in formal discussions with USDA-ARS leadership on the importance of refilling John Lydon's position of National Program Leader for Weed Science.
- Work with new APHIS PPQ leadership to reestablish federal funding for noxious weed control and eradication. The witchweed eradication program is the only funding that has survived.
- Hold a successful National Invasive Species Awareness Week (NISAW)

2013 NEWSS BUSINESS MEETING

February 6, 2013

Hilton Baltimore, Baltimore, MD

1) **Call to order – President**
   - President Toni DiTommaso called the business meeting to order at 5:09pm.
2) **Presidential Address**
   - Toni DiTommaso gave his Presidential address entitled "My Journey to the Presidency of the NEWSS" which gave a look back at his life across two continents and how his different experiences along the way led him to a career in Weed Science.
3) **Acceptance of the Minutes from the last Annual Business Meeting**
   - Brian Olson moved to accept the minutes and Jacob Barney seconded the motion. The motion passed unanimously.
4) **Necrology Report - Membership Chair**
   - Melissa Bravo mentioned the loss of several NEWSS members during the 2012 year which included Stan Pruss, Garry Schnappinger, John Lyndon, and John Ahrens. Greg Armel and Todd Mervosh addressed the audience on the life and impacts of Garry Schnappinger and John Ahrens, respectively.
5) **President's Comments (status of meeting, any updates)**
   - DiTommaso stated that the joint meeting with the WSSA had gone very well.
   - He addressed the fact that even though we did not have proceedings on hand this year as in the past (due to the joint meeting limitations) that these proceedings would be completed in the coming weeks and be available for sale.
6) **Treasurer Update**
   - Melissa Bravo reported that the society had a total net worth of $76,116 with a net gain of $11,536. In addition, the joint meeting with the WSSA would lead to a future surplus of funds that the society would need to determine how to appropriate.
   - Mark VanGessel provided Melissa Bravo a plaque for her service as Secretary-Treasurer of the society.
7) **Audit Committee Report**
   - The audit committee made up of Jim Steffel and Erin Hitchner reported their approval of the treasurer's report. Mark VanGessel moved to approve the audit committee's report with Brad Majek seconding the motion. The motion passed unanimously.
8) **Archives Committee Report**
   - Dan Kunkel reported that he had received the needed updates for the archives from Mark VanGessel last week.
9) **Awards Committee – Past-President**

   Garry Schnappinger Service Award
   - Mark VanGessel introduced Anne Schnappinger and her daughters Lee and Amy to dedicate the M. Garry Schnappinger Service Award. Anne gave a very gracious and heartfelt speech thanking the society for bestowing this honor on Garry. She gave the award to its first recipient Quintin Johnson.

   Awards presented at General Session are only mentioned
   - (a) Fellow- Bill Curran, Penn State University
   - (b) Award of Merit- Henry Wilson, Virginia Tech

(c) Outstanding Educator- Randy Prostak, UMASS
(d) Outstanding Researcher- 1) Jim Brosnan, University of Tennessee; 2) Richard Stalter, St. John's University
(e) Robert D. Sweet Outstanding Graduate Student(s)- Angela Post, Virginia Tech
(f) Garry Schnappinger Service Recognition- Quintin Johnson, University of Delaware
(g) Collegiate Weed Contest Winners (mentioned)

ii) Graduate Student Presentation awards - committee chair

- Mark VanGessel stated that they had 24 students competing in this year's contest.
  Am Student 2nd place: John Orlowski, University of Kentucky for the paper entitled "Alternate Use of PPO Herbicides in in High-Yield Soybean Management"
  Am Student 1st Place: Kelly Patches Penn State University for the paper entitled "Pokeweed Biology and Management in Pennsylvania Field Crops"
  Pm Student 2nd Place: Dan Tekiela, Virginia Tech for the paper entitled "Is Controlling Japanese Stiltgrass Worth it? Forest Understory Community Response."
  Pm Student 1st Place: Katherine Ghantous, UMASS, for the paper entitled "Effects of Timing and Frequency of Flame Cultivation Treatments on Dewberry"

iii) Research Poster contest - committee chair

- Quintin Johnson stated that they had 12 total posters in the poster contest this year.
  Student 2nd place: Rachel Atwell, NC State University, for the poster entitled "The Effects of Row Spacing and Population on Weed Competition in Organic Canola Production"
  Student 1st place: Kate Venner, Virginia Tech, for the poster entitled "Topramezone for Nimblewill and Groundivy Control in Cool Season Turf"

iv) Photo contest - committee chair

- Erin Hitchner made a plea for additional photos for next year's contest as few entries were provided for this year's event.
  2nd place: Farnaz Kordbacheh, University of Tehran, embryonic development and seed photos of velvetleaf
  1st place: Larissa Smith, Virginia Tech, *Miscanthus* X *Giganteus* seed

10) Old Business - NEWSS Endowment Fund – Greg Armel

- Greg Armel updated the membership on the approval of establishing an NEWSS endowment. He mentioned that the NEWSS board had approved the allocation of a $5000 from a budget surplus to help fund the endowment. He motioned to the membership to approve this allotment of funds. Brad Majek seconded the motion and the motion carried unanimously.

11) Officer Changeover and Presentation of the Gavel

- Toni DiTommaso thanked the outgoing members of the board and welcomed those new additions. Art Gover will be taking over for Darren Lycan as editor. Todd Mervosh will be filling the role of Public Relations Chair vacated by Javier Vargas. Erin Hitchner will be taking over the role of Treasurer from Melissa Bravo. John Orlowski took over the role of Graduate Student Representative from Adam Smith. Toni DiTommaso then transferred the office of President over the Dwight Lingenfelter who in turn provided Toni a plaque commemorating his service to the NEWSS.

12) Resolutions - committee chair

- Todd Mervosh stated that there were no new resolutions coming from the Resolutions committee.

13) New Committee Appointments

i) Resolutions Committee

ii) Nominating Committee - committee chair

(a) Election of the Vice President

- Rakesh Chandran of West Virginia University was the Nominating Committees choice for Vice-President. Dave Yarborough moved to close nominations for Vice-President and Dan Kunkel seconded the motion. The motion carried. Dwight asked for a vote on Rakesh Chandran as the new Vice-President and he won the post unanimously.

(b) Presidential Appointment (2) and Election (3) of the Nominating Committee

- Mark VanGessel will head the new Nominating Committee. Sudeep Matthew, Larissa Smith, Katherine Ghantous, and Bill Curran were all nominated to serve on the committee and all agreed to serve.

iii) Resolution Committee Presidential Appointments (Chair and members [2])

- Andy Senesac was appointed as Chair of the Resolutions Committee and Courtney Stokes and Jennifer D'Appollonio-Cote will serve as committee members.

14) Weed Contest Site and Date

- Dwight Lingenfelter stated that the next collegiate weed contest will be hosted by Monsanto in Monmouth, IL on July 24-25th, 2013. This contest will be a joint event between the NEWSS and NCWSS. The NEWSS will provide some travel funds to participating schools for making the long trek to Illinois for this event.

15) Next NEWSS Meeting Site

- Dwight Lingenfelter stated that the next NEWSS meeting would be held from January 6-9th, 2014 at the Sheraton Society Hill in Philadelphia, PA. Room rates will be $124/night with $17/night parking and free in room internet.

16) Presentation of the Executive Committee

- Dwight Lingenfelter presented the new NEWSS Executive committee to the membership.

17) Adjourn

- Mark VanGessel moved to close the business meeting and Hilary Sandler seconded the motion. The motion carried unanimously and the meeting ended at 6:30pm.

NEWSS PAST PRESIDENT

Gilbert H. Ahlgren	1947-49
Robert D. Sweet	1949-50
Howard L. Yowell	1950-51
Stephen M. Raleigh	1951-52
Charles E. Minarik	1952-53
Robert H. Beatty	1953-54
Albin O. Kuhn	1954-55
John Van Geluwe	1955-56
L. Danielson	1956-57
Charles L. Hovey	1957-58
Stanford N. Fertig	1958-59
Gordon Utter	1959-60
E. M. Rahn	1960-61
Lawrence Southwick	1961-62
Donald A. Shallock	1962-63
Anthony J. Tafuro	1963-64
Robert A. Peters	1964-65
Gideon D. Hill	1965-66
Richard D. Ilnicki	1966-67
John E. Gallagher	1967-68
John A. Meade	1968-69
Homer M. Lebaron	1969-70
John F. Ahrens	1970-71
George H. Bayer	1971-72
Arthur Bing	1972-73
Ralph Hansen	1973-74
Walter A. Gentner	1974-75
Henry P. Wilson	1975-76
Richard J. Marrese	1976-77
C. Edward Beste	1977-78
James D. Riggleman	1978-79
James V. Parochetti	1979-80
M. Garry Schnappinger	1980-81
Raymond B. Taylorson	1981-82
Stephan Dennis	1982-83
Thomas L. Watschke	1983-84
James C. Graham	1984-85
Russell R. Hahn	1985-86
Edward R. Higgins	1986-87
Maxwell L. McCormack	1987-88
Roy R. Johnson	1988-89
Stanley F. Gorski	1989-90
John B. Dobson	1990-91
Prasanta C. Bhowmik	1991-92
Stanley W. Pruss	1992-93
Ronald L. Ritter	1993-94
Wayne G. Wright	1994-95
Bradley A. Majek	1995-96
Thomas E. Vrabel	1996-97
Joseph C. Neal	1997-98
David B. Vitolo	1998-99
A. Richard Bonanno	1999-00
Brian D. Olson	2000-01
Jeffrey F. Derr	2001-02
David J. Mayonado	2002-03
D. Scott Glenn	2003-04
Robin R. Bellinder	2004-05
Timothy E. Dutt	2005-06
William S. Curran	2006-07
Renee Keese	2007-08
Jerry Baron	2008-09
David Yarborough	2009-10
Hilary Sandler	2010-11
Mark VanGessel	2011-12
Antonio DiTommaso	2012-13

NEWSS AWARD OF MERIT

1971	Gilbert H. Ahlgren	Rutgers University
	Homer Neville	L.I. Ag. & Tech, Farmingdale, NY
	Claude E. Phillips	University of Delaware
	M. S. Pridham	Cornell University
	Stephen A. Raleigh	Penn State University
1972	Robert Bell	University of Rhode Island
	Stuart Dunn	University of New Hampshire
	Alfred Fletcher	NJ State Dept. of Health
	Frank N. Hewetson	Penn State Fruit Research Lab.
	Madelene E. Pierce	Vassar College
	Collins Veatch	West Virginia University
	Howard L. Yowell	Esso Research Laboratory
1973	Moody F. Trevett	University of Maine
1974	Robert H. Beatty	Amchem Products, Inc.
	Arthur Hawkins	University of Connecticut
1975	Philip Gorlin	NY City Envir. Control Agency
	Herb Pass	CIBA-GEIGY Corp.
	Robert D. Sweet	Cornell University
1976	C. E. Langer	University of New Hampshire
	Charles E. Minarik	USDA-ARS
	Herb Pass	CIBA-GEIGY Corp.
1977	L. L. Danielson	US Dept. of Agriculture-ARS
	Madelene E. Pierce	Vassar College
	Lawrence Southwick	Dow Chemical Company
	John Stennis	USDI- Bureau of Fish & Wildlife
1978	*None Awarded*	
1979	Carl M. Monroe	Shell Chemical Company
	Charles Joseph Noll	Penn State University
	Jonas Vengris	University of Massachusetts
1980	Otis F. Curtis, Jr.	NY Ag. Experiment Station
	Theodore R. Flanagan	University of Vermont
	Oscar E. Shubert	Virginia University
1981	Dayton L. Klingman	USDA-ARS
	Hugh J. Murphy	University of Maine
	John Van Geluwe	CIBA-GEIGY Corp.
1982	Robert D. Shipman	Penn State University
1983	Arthur Bing	Cornell University
	William E. Chappel	Virginia Tech
	Barbara H. Emerson	Union Carbide Agricultural Prod.

1984	William H. Mitchell Roger S. Young	University of Delaware West Virginia University
1985	John A. Jagschitz	University of Rhode Island
1986	John R. Havis	University of Massachusetts
1987	*None Awarded*	
1988	J. Lincoln Pearson	University of Rhode Island
1989	Robert A. Peters	University of Connecticut
1990	Bryant L. Walworth	American Cyanamid Co.
1991	Don Warholic	Cornell University
1992	Robert Duell Richard Ilnicki William V. Welker	Rutgers University Rutgers University USDA-ARS
1993	*None Awarded*	
1994	John F. Ahrens John B. Dobson J. Ray Frank	CT Ag. Experiment Station American Cyanamid USDA-ARS/IR-4
1995	Francis J. Webb	University of Delaware
1996	Robert M. Devlin Wilber F. Evans Raymond B. Taylorson S. Wayne Bingham	University of Massachusetts Rhone-Poulenc Ag. Co. University of Rhode Island Virginia Tech
1997	Jean P. Cartier	Rhone-Poulenc Ag. Co.
1998	Stan Pruss Max McCormack, Jr.	Novartis Crop Protection University of Maine
1999	*None awarded*	
2000	Richard J. Marrese	Hoechst-NorAm
2001	Nathan L. Hartwig Edward R. Higgins	Penn State University Novartis Crop
2002	Garry Schnappinger	Syngenta Crop Protection
2003	*None Awarded*	
2004	C. Edward Beste James C. Graham	University of Maryland Monsanto
2005	Thomas L. Watschke	Penn State University
2006	Steve Dennis	Syngenta Crop Protection
2007	*None awarded*	
2008	Domingo Riego	Monsanto
2009	*None awarded*	

2010	Betty Marose	Marose Agricultural Consulting
2011	Scott Hagood	Virginia Tech
2012	Ronald Ritter	University of Maryland
2013	Henry Wilson	Virginia Tech

NEWSS FELLOW

1979	George H. Bayer Robert A. Peters Robert D. Sweet	Agway, Inc. University of Connecticut Cornell University
1980	John F. Ahrens John E. Gallagher Richard Ilnicki	CT Ag. Experiment Station Union Carbide Agric. Products Rutgers University
1981	Robert H. Beatty Arthur Bing John A. Meade	Amchem Products, Inc. Cornell University Rutgers University
1982	Walter A. Gentner Hugh J. Murphy	USDA-ARS University of Maine
1983	L. L. Danielson	USDA-ARS
1984	Barbara H. Emerson Henry P. Wilson	Union Carbide Agric. Products Virginia Tech
1985	*None Awarded*	
1986	Chiko Haramaki Dean L. Linscott	Penn State University USDA-ARS/Cornell University
1987	Gideon D. Hill Williams V. Welker	E. I. DuPont DeNemours US Dept. of Agric-ARS
1988	Wendell R. Mullison James V. Parochetti	Dow Chemical US Dept. of Agriculture-CSRS
1989	*None Awarded*	
1990	Robert M. Devlin	University of Massachusetts
1991	John (Jack) B. Dobson Robert D. Shipman	American Cyanamid Penn State University
1992	Gary Schnappinger	Ciba-Geigy Corp.
1993	Steve Dennis James Graham	Zeneca Ag. Products Monsanto
1994	Russell Hahn Maxwell McCormick	Cornell University University of Maine
1995	Richard Ashley Richard Marrese	University of Connecticut Hoechst-NorAm

1996	Roy R. Johnson Edward R. Higgins	Waldrum Specialist Inc. Ciba Crop Protection
1997	Raymond B. Taylorson Wayne G. Wright Stanley F. Gorski	UDSA-ARS DowElanco Ohio State University
1998	Prasanta Bhowmik	University of Massachusetts
1999	C. Edward Beste	University of Maryland
2000	J. Ray Frank Stanley W. Pruss	IR-4 Project Ciba Crop Protection
2001	Ronald L. Ritter	University of Maryland
2002	Bradley A. Majek Thomas L. Watschke	Rutgers University Penn State University
2003	Nathan L. Hartwig	Penn State University
2004	C. Benjamin Coffman Joseph C. Neal	USDA-ARS North Carolina State University
2005	David Vitolo	Syngenta Crop Protection
2006	A. Richard Bonnano Thomas Vrabel	University of Massachusetts Eco-Soil Systems, Central H.S.
2007	Larry Kuhns Brian Olsen	Penn State University Dow Agrosciences
2008	Jeff Derr	Virginia Tech
2009	David Mayonado Andrew Senesac	Monsanto Co. Cornell University
2010	Scott Glenn	University of Maryland
2011	Robin Bellinder	Cornell University
2012	Timothy Dutt	Monsanto/LAB Services
2013	William Curran	Penn State University

NEWSS OUTSTANDING RESEARCHER AWARD

1999	Garry Schnappinger	Novartis Crop Protection
2000	Prasanta C. Bhowmik	University of Massachusetts
2001	Robin Bellinder	Cornell University
2002	Jerry J. Baron	IR-4 Project, Rutgers University
2003	Arthur E. Gover	Penn State University
2004	Mark J. VanGessel	University of Delaware
2005	Bradley A. Majek	Rutgers University

2006	Grant Jordan	ACDS Research
2007	Peter Dernoeden	University of Maryland
2008	Shawn Askew	Virginia Tech
2009	Joseph Neal	North Carolina State University
2010	Todd Mervosh	CT Ag. Experiment Station
2011	Robert Richardson	North Carolina State University
2012	*None awarded*	
2013	James Brosnan Richard Stalter	University of Tennessee St. John's University

NEWSS OUTSTANDING EDUCATOR AWARD

1999	Douglas Goodale	SUNY Cobleskill
2000	Thomas L. Watschke	Penn State University
2001	C. Edward Beste	University of Maryland
2002	E. Scott Hagood	Virginia Tech University
2003	Andrew F. Senesac	Cornell University
2004	William S. Curran	Pennsylvania State University
2005	Antonio DiTommaso	Cornell University
2006	Russell Hahn	Cornell University
2007	Prasanta Bhowmik	University of Massachusetts
2008	Mike Fidanza	Penn State University
2009	Scott Glenn	University of Maryland
2010	Dwight Lingenfelter	Penn State University
2011	*None awarded*	
2012	*None awarded*	
2013	Randy Prostak	University of Massachusetts

M. GARRY SCHNAPPINGER SERVICE RECOGNITION AWARD

2013	Quintin Johnson	University of Delaware

NEWSS SERVICE RECOGNITION AWARD

2009	Thomas Hines	Virginia Tech
2010	Irene Tsontakis-Bradley	Cornell University

2011	Greg Breeden	University of Tennessee
	Robert Dickerson	Penn State University
2012	Sherilynn Novak	IR-4 Project

OUTSTANDING STUDENT PAPER CONTEST

1979	1	Bradley Majek	Cornell University
	2	Betty J. Hughes	Cornell University
1980	1	John Cardina	Penn State University
	2	Timothy Malefyt	Cornell University
1981	1	A. Douglas Brede	Penn State University
	2	Ann S. McCue	Cornell University
1982	1	Thomas C. Harris	University of Maryland
	2	Barbara J. Hook	University of Maryland
	HM	L. K. Thompson	Virginia Tech
	HM	Timothy Malefyt	Cornell University
1983	1	Anna M. Pennucci	University of Rhode Island
	2	Michael A. Ruizzo	Ohio State University
	HM	I. M. Detlefson	Rutgers University
1984	1	Robert S. Peregoy	University of Maryland
	2	Ralph E. DeGregorio	University of Connecticut
1985	1	Stephan Reiners	Ohio State University
	2	Erin Hynes	Penn State University
1986	1	Elizabeth Hirsh	University of Maryland
	2	Ralph E. DeGregorio	University of Connecticut
	2	Avraham Y. Teitz	Ohio State University
1987	1	Russell W. Wallace	Cornell University
	2	Daniel E. Edwards	Penn State University
	2	Frank J. Himmelstein	University of Massachusetts
1988	1	William K. Vencill	Virginia Tech
	2	Lewis K. Walker	Virginia Tech
	HM	Scott Guiser	Penn State University
	HM	Frank J. Himmelstein	University of Massachusetts
1989	1	Frank S. Rossi	Cornell University
	1	Amy E. Stowe	Cornell University
1990	1	William J. Chism	Virginia Tech
	2	Russell W. Wallace	Cornell University

1991	1	Elizabeth Maynard	Cornell University
	2	Daniel L. Kunkel	Cornell University
1992	1	J. DeCastro	Rutgers University
	2	Ted Blomgren	Cornell University
	3	Fred Katz	Rutgers University
1993	1	Eric D. Wilkens	Cornell University
	2	Henry C. Wetzel	University of Maryland
1994	1	Jed B. Colquhoun	Cornell University
	2	Eric D. Wilkins	Cornell University
1995	1	Sydha Salihu	Virginia Tech
	2	John A. Ackley	Virginia Tech
	HM	Jed B. Colquhoun	Cornell University
1996	1	Dwight Lingenfelter	Penn State University
	2	Mark Issacs	University of Delaware
	HM	Jed B. Colquhoun	Cornell University
1997	1	David Messersmith	Penn State University
	2	Sowmya Mitra	University of Massachusetts
	HM	Mark Issacs	University of Delaware
1998	1	Dan Poston	Virginia Tech
	2	Travis Frye	Penn State University
	3	David B. Lowe	Clemson University
1999	1	Hennen Cummings	North Carolina State University
	2	John Isgrigg	North Carolina State University
2000	1	Matthew Fagerness	North Carolina State University
	2	Steven King	Virginia Tech
	3	Gina Penny	North Carolina State University
2001	1	Robert Nurse	University of Guelph
	2	W. Andrew Bailey	Virginia Tech
	2	Steven King	Virginia Tech
2002	1	G. Michael Elston	University of Massachusetts
	2	Caren A. Judge	North Carolina State University
2003	1.	Matt Myers	Penn State University
	2	J. Scott McElroy	North Carolina State University
	3	Robert Nurse	Cornell University
2004	1	Whitnee L. Barker	Virginia Tech
	2	Caren A. Judge	North Carolina State University
	3	Erin R. Haramoto	University of Maine

2005	1	Jacob Barney	Cornell University
	2	Steven Mirsky	Penn State University
2006	1	Steven Mirsky	Penn State University
	1	Robert Shortell	Rutgers University
	2	Bryan Dillehay	Penn State University
2007	1	Bryan Dillehay	Penn State University
	2	John Willis	Virginia Tech
	3	Glenn Evans	Cornell University
2008	1	Glenn Evans	Cornell University
	2	Alex Putnam	University of Connecticut
	3	Angela Post	North Carolina State University
2009	1	Dustin Lewis	University of Tennessee
	2	Kristine Averill	Cornell University
	3	Angela Post	North Carolina State University
2010	1	Katherine Ghantous	University of Massachusetts
	2	Angela Post	Virginia Tech
	3	Matthew Cutulle	Virginia Tech
2011	1	Scott Wells	North Carolina State
	2	Brendan McNulty	Virginia Tech
2012	1	Kristine Averill	Penn State University
	2	Dan Tekiela	Virginia Tech
2013 (I)	1	Kelly Patches	Penn State University
	2	John Orlowski	University of Kentucky
2013 (II)	1	Katherine Ghantous	University of Massachusetts
	2	Dan Tekiela	Virginia Tech

DR. ROBERT D. SWEET OUTSTANDING GRADUATE STUDENT

2009	Jacob Barney (PhD)	Cornell University
2010	Matt Ryan (PhD)	Penn State University
	Katherine Ghantous (MS)	University of Massachusetts
2011	Kristine Averill (MS)	Cornell University
2012	Matthew Cutulle (PhD)	Virginia Tech
	Jose Vargas (MS)	University of Tennessee
2013	Angela Post (PhD)	Virginia Tech

NEWSS COLLEGIATE WEED CONTEST WINNERS

1983 - Wye Research Center, Maryland
- Graduate Team: University of Guelph
- Undergraduate Team: Penn State University
- Graduate Individual: Mike Donnelly, University of Guelph
- Undergraduate Individual: Bob Annet, University of Guelph

1984 Rutgers Research and Development Center, Bridgeton, New Jersey
- Graduate Team: University of Guelph
- Undergraduate Individual: D. Wright, University of Guelph
- Graduate Individual: N. Harker, University of Guelph

1985 Rohm and Haas, Spring House, Pennsylvania
- Graduate Team: University of Maryland
- Undergraduate Individual: Finlay Buchanan, University of Guelph
- Graduate Individual: David Vitolo, Rutgers University

1986 FMC, Princeton, New Jersey
- Graduate Team:
- Undergraduate Team: University of Guelph
- Graduate Individual: R. Jain, Virginia Tech
- Undergraduate Individual: Bill Litwin, University of Guelph

1987 DuPont, Newark, Delaware
- Graduate Team: University of Guelph
- Undergraduate Team: University of Guelph
- Graduate Individual: Lewis Walker, Virginia Tech
- Undergraduate Individual: Allen Eadie, University of Guelph

1988 Ciba-Geigy Corp., Hudson, New York
- Graduate Team: Virginia Tech
- Undergraduate Team: University of Guelph
- Graduate Individual: Carol Moseley, Virginia Tech
- Undergraduate Individual: Del Voight, Penn State University

1989 American Cyanamid, Princeton, New Jersey
- Graduate Team: Cornell University
- Undergraduate Team: SUNY Cobleskill
- Graduate Individual: Paul Stachowski, Cornell University
- Undergraduate Individual: Anita Dielman, University of Guelph

1990 Agway Farm Research Center, Tully, New York
- Graduate Team: Virginia Tech
- Undergraduate Team: SUNY Cobleskill
- Graduate Individual: Brian Manley, Virginia Tech
- Undergraduate Individual: Dwight Lingenfelter, Penn State University

1991 Rutgers University, New Brunswick, New Jersey
- Graduate Team: Virginia Tech
- Undergraduate Team: University of Guelph

Graduate Individual: Carol Moseley, Virginia Tech
Undergraduate Individual: Tim Borro, University of Guelph

1992 Ridgetown College, Ridgetown, Ontario, Canada
Graduate Team: Michigan State University
Undergraduate Team: Ohio State
Graduate Individual: Troy Bauer, Michigan State University
Undergraduate Individual: Jeff Stackler, Ohio State University

1993 Virginia Tech, Blacksburg, Virginia
Graduate Team: Virginia Tech
Undergraduate Team: SUNY Cobleskill
Graduate Individual: Brian Manley, Virginia Tech
Undergraduate Individual: Brian Cook, University of Guelph

1994 Lower Eastern Shore Research and Education Center, Salisbury, Maryland
Graduate Team: Virginia Tech
Undergraduate Team: University of Guelph
Graduate Individual: Brian Manley, Virginia Tech
Undergraduate Individual: Robert Maloney, University of Guelph

1995 Thompson Vegetable Research Farm, Freeville, New York
Graduate Team: Virginia Tech
Undergraduate Team: University of Guelph
Graduate Individual: Dwight Lingenfelter, Penn State University
Undergraduate Individual: Jimmy Summerlin, North Carolina State Univ.

1996 Penn State Agronomy Farm, Rock Springs, Pennsylvania
Graduate Team: Michigan State University
Undergraduate Team: SUNY, Cobleskill
Graduate Individual: John Isgrigg, North Carolina State University
Undergraduate Individual: Mark Brock, University of Guelph

1997 North Carolina State University, Raleigh, North Carolina
Graduate Team: Michigan State University
Undergraduate Team: University of Guelph
Graduate Individual: Brett Thorpe, Michigan State University

1998 University of Delaware, Georgetown, Delaware
Graduate Team: Virginia Tech
Undergraduate Team: University of Guelph
Graduate Individual: Shawn Askew, North Carolina State University
Undergraduate Individual: Kevin Ego, University of Guelph

1999 Virginia Tech, Blacksburg, Virginia
Graduate Team: North Carolina State University
Undergraduate Team: Nova Scotia Agricultural College
Graduate Individual: Rob Richardson, Virginia Tech
Undergraduate Individual: Keith Burnell, North Carolina State University

2000 University of Guelph, Guelph, Ontario, Canada
Graduate Team: Virginia Tech
Undergraduate Team: Ohio State University

Graduate Individual: Shawn Askew, North Carolina State University
Undergraduate Individual: Luke Case, Ohio State University

2001 University of Connecticut, Storrs, Connecticut
Graduate Team: North Carolina State University
Undergraduate Team: Penn State University
Graduate Individual: Matt Myers, Penn State University
Undergraduate Individual: Shawn Heinbaugh, Penn State University

2002 ACDS Research Facility, North Rose, New York
Graduate Team: North Carolina State University
Undergraduate Team: North Carolina State University
Graduate Individual: Scott McElroy, North Carolina State University
Undergraduate Individual: Sarah Hans, North Carolina State University

2003 Syngenta Crop Protection, Eastern Region Technical Center, Hudson, NY
Graduate Team: North Carolina State University
Undergraduate Team: University of Guelph
Graduate Individual: Andrew MacRae, North Carolina State University
Undergraduate Individual: Jonathan Kapwyk, University of Guelph

2004 North Carolina University, Raleigh, NC
Graduate Team: North Carolina State University
Undergraduate Team: University of Guelph
Graduate Individual: John Willis, Virginia Tech
Undergraduate Individual: Jenny English, University of Guelph

2005 Pennsylvania State University, Landisville, PA
Graduate Team: North Carolina State University
Undergraduate Team: University of Guelph
Graduate Individual: John Willis, Virginia Tech
Undergraduate Individual: Gerard Pynenborg, University of Guelph

2006 DuPont Crop Protection, Stine Haskell Research Center, Newark, Delaware
Graduate Team: North Carolina State University
Undergraduate Team: University of Guelph
Graduate Individual: Virender Kumar, Cornell University
Undergraduate Individual: Adam Pfeffer, University of Guelph

2007 Virginia Tech, Blacksburg, Virginia
Graduate Team: North Carolina State University
Undergraduate Team: University of Guelph
Graduate Individual: George Place, North Carolina State University
Undergraduate Individual: Craig Reid, University of Guelph

2008 University of Delaware, Georgetown, Delaware
Graduate Team: Penn State University
Undergraduate Team: University of Guelph
Graduate Individual: Matt Ryan, Penn State University
Undergraduate Individual: Blair Scott, University of Guelph

2009 ABG Ag Services, Sheridan, Indiana (joint contest with the NCWSS)
Graduate Team: Penn State University

Undergraduate Team: University of Guelph
Graduate Individual: Angela Post, Cornell University
Undergraduate Individual: Andrew Reid, University of Guelph

2010 Cornell University, Freeville, New York
Graduate Team: Michigan State University
Undergraduate Team: University of Guelph
Graduate Individual: Jason Parish, Ohio State University
Undergraduate Individual: Cory Chelko, Penn State University

2011 University of Tennessee, Knoxville, Tennessee (Weed Olympics)
Graduate-NEWSS: North Carolina State University
Graduate-Overall: Purdue University
Undergraduate-NEWSS and Overall: University of Guelph
Graduate-NEWSS: Dustin Lewis, North Carolina State University
Graduate-Overall: Jason Parrish, The Ohio State University
Undergraduate-NEWSS and Overall: Dan Tekiela, Virginia Tech

2012 Rutgers University Snyder Farm, Pittstown, New Jersey
Graduate Team: Virginia Tech
Undergraduate Team: University of Guelph
Graduate Individual: Alex Lindsey, Ohio State University
Undergraduate Individual: Dennis Cooper, University of Guelph

2013 Monsanto Research and Learning Center, Monmouth, Illinois (with NCWSS)
Graduate Team-NEWSS: North Carolina State University
Undergraduate Team-NEWSS: Guelph University
Graduate Individual-NEWSS: Sushila Chaudhari, North Carolina State U.
Undergraduate Individual-NEWSS: Johanna Burrows, Univ. of Guelph

RESEARCH POSTER AWARDS

1983 1 Herbicide Impregnated Fertilizer of Weed Control in No-Tillage Corn. R. Uruatowski and W. H. Mitchell, Univ. of Delaware, Newark.

2 Effect of Wiper Application of Several Herbicides and Cutting on Black Chokeberry. D. E. Yarborough and A. A. Ismail, Univ. of Maine, Orono.

HM Corn Chamomile Control in Winter Wheat. R. R. Hahn, Cornell Univ., Ithaca, New York and P. W. Kanouse, New York State Cooperative Extension, Mt. Morris.

1984 1 Herbicide Programs and Tillage Systems for Cabbage. R. R. Bellinder, Virginia Tech, Blacksburg, and T. E. Hines and H. P. Wilson, Virginia Truck and Ornamental Research Station, Painter.

2 Triazine Resistant Weeds in New York State. R. R. Hahn, Cornell Univ., Ithaca, NY.

HM A Roller for Applying Herbicides at Ground Level. W. V. Welker and D. L. Peterson, USDA-ARS, Kearneysville, WV.

1985	1.	No-Tillage Cropping Systems in a Crown Vetch Living Mulch. N. L. Hartwig, Penn State Univ., University Park.
	2.	Anesthetic Release of Dormancy in *Amaranthus retroflexus* Seeds. R. B. Taylorson, USDA-ARS, Beltsville, MD and K. Hanyadi, Univ. of Agricultural Science, Keszthely, Hungary.
	2	Triazine Resistant Weed Survey in Maryland. B. H. Marose, University of Maryland, College Park.
	HM	Wild Proso Millet in New York State. R. R. Hahn, Cornell Univ., Ithaca, NY.
1986	1	Discharge Rate of Metolachlor from Slow Release Tablets. S. F. Gorski, M. K. Wertz and S. Refiners, Ohio State Univ., Columbus.
	2	Glyphosate and Wildlife Habitat in Maine. D.Santillo, University of Maine, Orono.
1987	1	Mycorrhiza and Transfer of Glyphosate Between Plants. M. A. Kaps and L. J. Kuhns, Penn State University, University Park.
	2	Redroot Pigweed Competition Study in No-Till Potatoes. R. W. Wallace, R. R. Bellinder, and D. T. Warholic, Cornell University, Ithaca, NY.
1988	1	Growth Suppression of Peach Trees With Competition. W.V. Welker and D.M. Glenn, USDA-ARS, Kearneysville, WV.
	2	Smooth Bedstraw Control in Pastures and Hayfields. R. R. Hahn, Cornell University, Ithaca, NY.
1989	1	Burcucumber Responses to Sulfonylurea Herbicides. H. P. Wilson and T. E. Hines, Virginia Tech, Painter, VA.
	2	Water Conservation in the Orchard Environment Through Management. W.V. Welker, Jr., USDA-ARS Appalachian Fruit Research Station, Kearneysville, WV.
1990	1	Reduced Rates of Postemergence Soybean Herbicides. E. Prostko, J. A. Meade, and J. Ingerson-Mahar, Rutgers Coop. Ext. Mt. Holly, NJ.
	2	The Tolerance of Fraxinus, Juglans, and Quercus Seedings to Imazaquin and Imazethapyr. L. J. Kuhns and J. Loose, Penn State University, University Park.
1991	1	Johnsongrass Recovery from Sulfonylurea Herbicides. T. E. Hines and H. P. Wilson, Virginia Tech, Painter.
	2	Growth Response to Young Peach Trees to Competition With Several Grass Species. W. V. Welker and D. M. Glenn, USDA-ARS, Appalachian Fruit Research Station, Kearneysville, WV.
1992	1	Teaching Weed Identification with Videotape. B. Marose, N. Anderson, L. Kauffman-Alfera, and T. Patten, University of Maryland, College Park.
	2	Biological Control of Annual Bluegrass (*Poa annua* L. *reptans)* with *Xanthomonas campestris* (MYX-7148) Under Field Conditions. N. D. Webber and J. C. Neal, Cornell University, Ithaca, NY.

1993	1	Development of an Identification Manual for Weeds of the Northeastern United States. R H. Uva and J. C. Neal, Cornell University, Ithaca, NY.
	2	Optimum Time of Cultivation for Weed Control in Corn. J. Mt. Pleasant, R. Burt and J. Frisch, Cornell University, Ithaca, NY.
1994	1	Herbicide Contaminant Injury Symptoms on Greenhouse Grown Poinsettia and Geranium. M. Macksel and A. Senesac, Long Island Horticultural Research Lab, Riverhead, NY and J. Neal, Cornell University, Ithaca, NY.
	2	Mow-kill Regulation of Winter Cereals Grown for Spring No-till Crop Production. E. D. Wilkins and R. R. Bellinder, Cornell Univ., Ithaca, NY.
1995	1	A Comparison of Broadleaf and Blackseed Plantains Identification and Control. J.C. Neal and C.C. Morse, Cornell University, Ithaca, NY.
	2	Using the Economic Threshold Concept as a Determinant for Velvetleaf Control in Field Corn. E.L. Werner and W.S. Curran, Penn State University, University Park.
1996	1	Preemergence and Postemergence Weed Management in 38 and 76 cm Corn. C. B. Coffman, USDA-ARS, Beltsville, MD.
	2	Common Cocklebur Response to Chlorimuron and Imazaquin. B.S. Manley, H.P. Wilson and T.E. Hines, Virginia Tech, Blacksburg, VA.
1997		*None Awarded*
1998	1	Weed Control Studies with *Rorippa sylvestris*. L.J. Kuhns and T. Harpster, Penn State University, University Park.
	2	Postemergence Selectivity and Safety of Isoxaflutole in Cool-season Turfgrass. P.C. Bhowmik and J.A. Drohen, Univ. of Massachusetts, Amherst.
1999	1	Winter Squash Cultivars Differ in Response to Weed Competition. E. T. Maynard, Purdue University, Hammond, IN.
	2	Effectiveness of Row Spacing, Herbicide Rate, and Application Method on Harvest Efficiency of Lima Beans. S. Sankula, M.J. VanGessel, W.E. Kee, and J.L. Glancey, University of Delaware, Georgetown.
2000	1	Weed Control and Nutrient Release With Composted Poultry Litter Mulch in a Peach Orchard. P.L. Preusch, Hood College, Frederick, MD; and T. J. Tworkoski, USDA-ARS, Kearneysville, WV.
	2	The Effect of Total Postemergence Herbicide Timings on Corn Yield. D.B. Vitolo, C. Pearson, M.G. Schnappinger, and R. Schmenk, Novartis Crop Protection, Hudson, NY.
	2	Pollen Transport from Genetically Modified Corn. J.M. Jemison and M. Vayda, University of Maine, Orono, ME.
2001	1	Evaluation of Methyl Bromide Alternatives for Yellow Nutsedge Control in Plasticulture Tomato. W. A. Bailey, H. P. Wilson, and T. E. Hines, Virginia Tech, Painter.

	2	Evaluation of Alternative Control Methods for Annual Ryegrass in Typical Virginia Crop Rotations. S. R. King and E. S. Hagood, Virginia Tech, Blacksburg.
2002	1	Effectiveness of Mesotrione to Control Weeds in Sweet Corn. J. M. Jemison, Jr. and A. Nejako, University of Maine, Orono.
	2	Flufenacet plus metribuzin for italian ryegrass control in Virginia wheat. W. A. Bailey, H. P. Wilson, and T. E. Hines, Virginia Tech, Painter.
2003	1	Comparison of Two Methods to Estimate Weed Populations in Field-scale Agricultural Research. R.D. Stout, M.G. Burton, and H.M. Linker, North Carolina State University, Raleigh.
	2	Diquat plus Glyphosate for Rapid-symptom Vegetation Control in Turf. W.L. Barker, S.D. Askew, J.B. Beam, Virginia Tech, Blacksburg; and D.C. Riego, Monsanto Co., Carmel, IN.
2004	1	Biology of the Invasive Plant Pale Swallow-wort. L. Smith, S. Greipsson, and A. DiTommaso, Cornell University, Ithaca, NY.
	2	Evaluating Perennial Groundcovers for Weed Suppression: Roadside Trials and Demonstrations. A. Senesac, I. Tsontakis-Bradley, J. Allaire, and L. Weston, Cornell University, Ithaca, NY.
2005	1	Cover Crop Management Impacts on the Weed Seed Predator, *Harpalus rufipes*. A. Shearin, S.C. Reberg-Horton, E. Gallandt, and F. Drummond, University of Maine, Orono.
	2	Carfentrazone, Quinclorac, and Trifloxysulfuron Effects on Seeded Bermudagrass Establishment and Crabgrass Control. J. Willis, D.B. Ricker, and S.D. Askew, Virginia Tech, Blacksburg.
2006	1	Mesotrione for Preemergence Broadleaf Weed Control in Turf. D. Ricker, J. Willis, and S. Askew, Virginia Tech, Blacksburg.
	2	Using a Wet-blade Mower for Pest Control, Fertility, and Growth Retardation in Fine Turfgrass. J. Willis and S.D. Askew, Virginia Tech, Blacksburg.
2007	1	Effects of Emergence Periodicity on Growth and Fecundity of Horseweed. J. Dauer, B.A. Scott, M.J. VanGessel, and D.A. Mortensen. Penn State University, University Park.
	2	Vascular Weed Control in Container Production Using Selected Non-chemical Top-dress Treatments. A. Burtt. University of Vermont, Burlington.
2008	1	Evaluation of the Impact of an Adventitious Herbivore on an Invasive Plant, Yellow Toadflax, in Colorado USA. J.F. Egan and R.E. Irwin. Penn State University, University Park.
	1	Organic Weed Management: What the Farmers Think. M.R. Ryan, D.A. Mortensen, D.O. Wilson, and P.R. Hepperly. Penn State University, University Park.

2009	1	Turfgrass Response to Herbicide-treated Irrigation Water. R.L. Roten, R.J. Richardson, and A.P. Gardner. North Carolina State University, Raleigh.
	2	Response of Cranberry Vines to Hand-held Flame Cultivators - Initial Year Evaluation. K.M. Ghantous, H.A. Sandler, and P. Jeranyama. University of Massachusetts, Amherst.
2010	1	Comparison of Genetic Diversity of Weedy and Domesticated Populations in Genus *Cichorium*. T. Zavada. University of Massachusetts, Boston.
	2	Use of Mesotrione for Annual Bluegrass Control at Kentucky Bluegrass Establishment. K. Venner, S. Hart, and C. Mansue. Rutgers University, New Brunswick, NJ.
	3	The Effects of Herbicide Mixtures with Brassica meal on Weed Control and Yield of Strawberry. J. Cummins, G. Armel, C. Sams, D. Deyton, and J. Vargas. University of Tennessee, Knoxville.
2011	1	Length of Residual Annual Bluegrass Control of Mesotrione Relative to other Preemergence Weed Control Products. S. Hart, C.J. Manuse, and K.A. Venner. Rutgers University, New Brunswick, NJ.
	2	Improving Survey Methods for the Detection of Giant Hogweed in Pennsylvania. I.D. Bowers, M.A. Bravo, and J. Zoschg. Penn State University, University Park.
	3	Influence of Nitrogen, Plant Growth Regulators, and Ferrous Sulfate on Annual Bluegrass Populations. K.M. Han and J.E. Kaminski. Penn State University, University Park.
2012	1	Hairy Vetch Seedbank Persistence as Influenced by Mechanical Scarification and Soil Depth. B.C. Crockett, S. Mirsky, and W.S. Curran, Pennsylvania State University, University Park.
	2	Effects of Methiozolin Rates on Creeping Bentgrass and Annual Bluegrass Root Growth. K.A. Venner, S. Askew, and S. Koo, Virginia Tech, Blacksburg.
2013	1	Topramezone for Nimblewill and Groundivy Control in Cool-season Turf. K. Venner, A.N. Smith, S.D. Askew, Virginia Tech, Blacksburg; K. Miller, BASF, Richmond, VA.
	2	The Effects of Row Spacing and Population on Weed Competition in Organic Canola Production. R.A. Atwell, North Carolina State University, Raleigh.

INNOVATOR OF THE YEAR

1986	Nathan Hartwig	Penn State University
1987	Thomas Welker	USDA-ARS
1988	*None Awarded*	
1989	John E. Waldrum	Union Carbide Agric. Products
1990	*None Awarded*	

1991	Thomas L. Watschke	Penn State University
1992	E. Scott Hagood	Virginia Tech
	Ronald L. Ritter	University of Maryland
1993	*None Awarded*	
1994	George Hamilton	Penn State University
1995	Kent D. Redding	DowElanco
1996	James Orr	Asplundh Tree Expert Co.
1997	George Hamilton	Penn State University
1998	*None Awarded*	
1999	*Award Discontinued*	

OUTSTANDING APPLIED RESEARCH IN FOOD AND FEED CROPS

1991	Russell R. Hahn	Cornell University
1992	Henry P. Wilson	Virginia Tech
1993	*None Awarded*	
1994	Robin Bellinder	Cornell University
1995	*None Awarded*	
1996	E. Scott Hagood	Virginia Tech
1997	Ronald L. Ritter	University of Maryland
1998	*None Awarded*	
1999	*Award Discontinued*	

OUTSTANDING APPLIED RESEARCH IN TURF, ORNAMENTALS, AND VEGETATION MANAGEMENT

1991	Wayne Bingham	Virginia Tech
1992	John F. Ahrens	CT Agricultural Experiment Station
1993	Joseph C. Neal	Cornell University
1994	Prasanta C. Bhowmik	University of Massachusetts
1995	Andrew F. Senesac	Long Island Horticultural Research Lab
1996	Larry J. Kuhns	Penn State University
197	Jeffrey F. Derr	Virginia Tech
1998	*None Awarded*	
1999	*Award Discontinued*	

OUTSTANDING PAPER AWARDS

1954	Studies on Entry of 2,4-D into Leaves - J. N. Yeatman, J. W. Brown, J. A. Thorne and J. R. Conover, Camp Detrick, Frederick, MD
1954	The Effect of Soil Organic Matter Levels on Several Herbicides - S. L. Dallyn, Long Island Vegetable Research Farm, Riverhead, NY
1954	Experimental Use of Herbicides Impregnated on Clay Granules for Control of Weeds in Certain Vegetable Crops - L. L. Danielson, Virginia Truck Expt. Station, Norfolk, VA
1954	Cultural vs. Chemical Weed Control in Soybeans - W. E. Chappell, Virginia Polytechnic Institute, Blacksburg, VA
1954	Public Health Significance of Ragweed Control Demonstrated in Detroit - J. H. Ruskin, Department of Health, Detroit, MI
1955	A Comparison of MCP and 2,4-D for Weed Control in Forage Legumes - M. M. Schreiber, Cornell Univ., Ithaca, NY
1956	*None Awarded*
1957	Herbicidal Effectiveness of 2,4-D, MCPB, Neburon and Others as Measured by Weed Control and Yields of Seedling Alfalfa and Birdsfoot Trefoil - A. J. Kerkin and R. A. Peters, Univ. of Connecticut, Storrs
1957	Progress Report #4 - Effects of Certain Common Brush Control Techniques and Material on Game Food and Cover on a Power Line Right-of-Way - W. C. Bramble, W. R. Byrnes, and D. P. Worley, Penn State Univ., University Park
1958	Effects of 2,4-D on Turnips - C. M. Switzer, Ontario Agricultural College, Guelph, Canada
1958	Ragweed Free Areas in Quebec and the Maritimes - E. E. Compagna, Universite Laval at Ste-Anne-de-la-Pocatiere, Quebec, Canada
1959	Yields of Legume-Forage Grass Mixtures as Affected by Several Herbicides Applied Alone or in a Combination During Establishment - W. G. Wells and R. A. Peters, Univ. of Connecticut, Storrs
1959	Influence of Soil Moisture on Activity of EPTC, CDEC and CIPC - J. R. Havis, R. L. Ticknor and P. F. Boblua, Univ. of Massachusetts, Amherst
1960	The Influence of Cultivation on Corn Yields When Weeds are Controlled by Herbicides - W. F. Meggitt, Rutgers Univ., New Brunswick, NJ
1961	Preliminary Investigation of a Growth Inhibitor Found in Yellow Foxtail *(Setaria glauca* L.) - H. C. Yokum, M. J. Jutras, and R. A. Peters, Univ. of Connecticut, Storrs
1962	The Effects of Chemical and Cultural Treatment on the Survival of Rhizomes and on the Yield of Underground Food Reserves of Quackgrass - H. M. LeBaron and S. N. Fertig, Cornell Univ., Ithaca, NY
1962	Observations on Distribution and Control of Eurasian Watermilfoil in Chesapeake Bay, 1961 - V. D. Stotts and C. R. Gillette, Annapolis, MD

1963 The Relation of Certain Environmental Conditions to the Effectiveness of DNBP of Post-Emergence Weed Control in Peas - G. R. Hamilton and E. M. Rahn, Univ. of Delaware, Newark

1963 The Influence of Soil Surface and Granular Carrier Moisture on the Activity of EPTC - J. C. Cialone and R. D. Sweet, Cornell Univ., Ithaca, NY

1963 The Determination of Residues of Kuron in Birdsfoot Trefoil and Grasses - M. G. Merkle and S. N. Fertig, Cornell Univ., Ithaca, NY

1964 Control of Riparian Vegetation with Phenoxy Herbicides and the Effect on Streamflow Quality - I. C. Reigner, USDA-Forest Service, New Lisbon, NJ; W. E. Sopper, Penn State Univ., University Park; and R. R. Johnson, Amchem Products, Inc., Ambler, PA

1964 EPTC Incorporation by Band Placement and Standard Methods in Establishment of Birdsfoot Trefoil - D. L. Linscott and R. D. Hagin, Cornell Univ., Ithaca, NY

1965 1 Corn Chamomile *(Anthemis arvensis* L.) Responses to Some Benzoic Acid Derivatives - Barbara M. Metzger, Judy K. Baldwin and R. D. Ilnicki, Rutgers Univ., New Brunswick, NJ

2 The Physical Properties of Viscous Sprays for Reduction of Herbicide Drift - J. W. Suggitt, The Hydro-Electric Power Commission of Ontario, Canada

1966 1 Weed Control Under Clear Plastic Mulch - Carl Bucholz, Cornell Univ., Ithaca, NY

2 A Chemical Team For Aerial Brush Control on Right-of-Way - B. C. Byrd and C. A. Reimer, Dow Chemical Co

1967 1 Influence of Time of Seeding on the Effectiveness of Several Herbicides Used for Establishing an Alfalfa-Bromegrass Mixture - R. T. Leanard and R. C. Wakefield, Univ. of New Hampshire, Durham

2 Weed Competition in Soybeans - L. E. Wheetley and R. H. Cole, Univ. of Delaware, Newark

1968 *None Awarded*

1969 1 Weed and Crop Responses in Cucumbers and Watermelons - H. P. Wilson and R. L. Waterfield, Virginia Truck and Orn. Res. Sta., Painter

2 Effect of Several Combinations of Herbicides on the Weight and Development of Midway Strawberry Plants in the Greenhouse - O. E. Schubert, West Virginia Univ., Morgantown

1970 1 Effects of RH-315 on Quackgrass and Established Alfalfa - W. B. Duke, Cornell Univ., Ithaca, NY

1971 1 Activity of Nitralin, Trifluralin and ER-5461 on Transplant Tomato and Eggplant - D. E. Broaden and J. C. Cialone, Rutgers Univ., New Brunswick, NJ

	2	Field Investigations of the Activities of Several Herbicides for the Control of Yellow Nutsedge - H. P. Wilson, R. L. Waterfield, Jr., and C. P. Savage, Jr., Virginia Truck and Orn. Res. Sta., Painter
1972	1	Study of Organisms Living in the Heated Effluent of a Power Plant - M. E. Pierce, Vassar College and D. Allessandrello, Marist College
	2	Effect of Pre-treatment Environment on Herbicide Response and Morphological Variation of Three Species - A. R. Templeton and W. Hurtt, USDA-ARS, Fort Detrick, MD
1973	1	A Simple Method of Expressing the Relative Efficacy of Plant Growth Regulators - A. R. Templeton and W. Hurtt, USDA-ARS, Fort Detrick, MD
	2	Agronomic Factors Influencing the Effectiveness of Glyphosate for Quackgrass Control –F. E. Brockman, W. B. Duke, and J. F. Hunt, Cornell Univ., Ithaca, NY
1974	1	Weed Control in Peach Nurseries - O. F. Curtis, Cornell Univ., Ithaca, NY
	2	Persistence of Napropamide and U-267 in a Sandy Loam Soil - R. C. Henne, Campbell Institute for Agr. Res., Napoleon, OH
1975	1	Control of Jimsonweed and Three Broadleaf Weeds in Soybeans - J. V. Parochetti, Univ. of Maryland, College Park
	HM	The Influence of Norflurazon on Chlorophyll Content and Growth of *Potomogeton pectinatus* - R. M. Devlin and S. J. Karcyzk, Univ. of Massachusetts, East Wareham
	HM	Germination, Growth, and Flowering of Shepherdspurse - E. K. Stillwell and R. D. Sweet, Cornell Univ., Ithaca, NY
1976	1	Top Growth and Root Response of Red Fescue to Growth Retardants - S. L. Fales, A. P. Nielson and R. C. Wakefield, Univ. of Rhode Island, Kingston
	HM	Selective Control of *Poa annua* in Kentucky Bluegrass - P. J. Jacquemin, O. M. Scott and Sons, and P. R. Henderlong, Ohio State Univ., Columbus
	HM	Effects of DCPA on Growth of Dodder - L. L. Danielson, USDA ARS, Beltsville, MD
1977	1	The Effects of Stress on Stand and Yield of Metribuzin Treated Tomato Plants - E. H. Nelson and R. A. Ashley, Univ. of Connecticut, Storrs
	HM	The Influence of Growth Regulators on the Absorption of Mineral Elements - R. M. Devlin and S. J. Karcyzk, Univ. of Massachusetts, East Wareham.
	HM	Quantification of S-triazine Losses in Surface Runoff: A Summary - J. K. Hall, Penn State Univ., University Park
1978	1	Annual Weedy Grass Competition in Field Corn - Jonas Vengris, Univ. of Massachusetts, Amherst
	HM	Metribuzin Utilization with Transplanted Tomatoes - R. C. Henne, Campbell Institute of Agr. Res., Napoleon, OH

1979 1 Herbicides for Ground Cover Plantings - J. F. Ahrens, Connecticut Agric. Expt. Station, Windsor

2 Weed Control Systems in Transplanted Tomatoes - R. C. Henne, Campbell Institute of Agr. Res. Napoleon, OH

1980 1 Integrated Weed Control Programs for Carrots and Tomatoes - R. C. Henne and T. L. Poulson, Campbell Institute of Agr. Res. Napoleon, OH

2 Suppression of Crownvetch for No-Tillage Corn - J. Carina and N. L. Hartwig, Penn State Univ., University Park

HM Effect of Planting Equipment and Time of Application on Injury to No-tillage Corn from Pendimethalin-Triazine Mixtures - N. L. Hartwig, Penn State Univ., University Park

1981 1 Weed Control in Cucumbers in Northwest Ohio - R. C. Henne and T. L. Poulson, Campbell Institute of Agr. Res. Napoleon, OH

2 Prostrate Spurge Control in Turfgrass Using Herbicides - J. A. Jagschitz, Univ. of Rhode Island, Kingston

HM Some Ecological Observations of Hempstead Plains, Long Island - R. Stalter, St. John's Univ., Jamaica, NY

1982 1 Differential Growth Responses to Temperature Between Two Biotypes of *Chenopodium album* - P. C. Bhowmik, Univ. of Massachusetts, Amherst

2 Chemical Control of Spurge and Other Broadleaf Weeds in Turfgrass - J. S. Ebdon and J. A. Jagschitz, Univ. of Rhode Island, Kingston

HM Influence of Norflurazon on the Light Activation of Oxyfluorfen - R. M. Devlin, S. J. Karczmarczyk, I. I. Zbiec and C. N. Saras, Univ. of Massachusetts, East Wareham

HM Analysis of Weed Control Components for Conventional, Wide-row Soybeans in Delaware - D. K. Regehr, Univ. of Delaware, Newark

1983 1 Comparisons of Non-Selective Herbicides for Reduced Tillage Systems - R. R. Bellinder, Virginia Tech, Blacksburg and H. P. Wilson, Virginia Truck and Orn. Res. Station, Painter

2 The Plant Communities Along the Long Island Expressway, Long Island, New York - R. Stalter, St. John's Univ., Jamaica, NY

HM Effect of Morning, Midday and Evening Applications on Control of Large Crabgrass by Several Postemergence Herbicides - B. G. Ennis and R.A. Ashley, Univ. of Connecticut, Storrs

1984 1 Pre-transplant Oxyfluorfen for Cabbage - J. R. Teasdale, USDA-ARS, Beltsville, MD

2 Herbicide Programs and Tillage Systems for Cabbage - R. R. Bellinder, Virginia Tech, Blacksburg and T. E. Hines and H. P. Wilson, Virginia Truck and Orn. Res. Station, Painter

1985	1	Peach Response to Several Postemergence Translocated Herbicides - B. A. Majek, Rutgers Univ., Bridgeton, NJ
1986	1	Influence of Mefluidide Timing and Rate on *Poa annua* Quality Under Golf Course Conditions - R. J. Cooper, Univ. of Massachusetts, Amherst; K. J. Karriok, Univ. of Georgia, Athens, and P. R. Henderlong and J. R. Street, Ohio State Univ., Columbus
	2	The Small Mammal Community in a Glyphosate Conifer Release Treatment in Maine - P. D'Anieri, Virginia Tech, Blacksburg; M. L. McCormack, Jr., Univ. of Maine, Orono; and D. M. Leslie, Oklahoma State Univ., Stillwater
	HM	Field Evaluation of a Proposed IPM Approach for Weed Control in Potatoes - D. P. Kain and J. B. Sieczka, Cornell Univ., Long Island Horticultural Research Laboratory, Riverhead, NY and R. D. Sweet, Cornell Univ., Ithaca, NY
1987		*None Awarded*
1988	1	Bentazon and Bentazon-MCPB Tank-mixes for Weed Control in English Pea - G. A. Porter, Univ. of Maine, Orono; A. Ashley, Univ. of Connecticut, Storrs; R. R. Bellinder and D. T. Warholic, Cornell Univ., Ithaca, NY; M. P. Mascianica, BASF Corp., Parsippany, NJ; and L. S. Morrow, Univ. of Maine, Orono
	2	Effects of Herbicide Residues on Germination and Early Survival of Red Oak Acorns - R. D. Shipman and T. J. Prunty, Penn State Univ., University Park
	2	Watershed Losses of Triclopyr after Aerial Application to Release Spruce Fir - C. T. Smith, Univ. of New Hampshire, Durham and M. L. McCormack, Jr., Univ. of Maine, Orono
1989		*None Awarded*
1990		*None Awarded*
1991		*Award Discontinued*

NEWSS MEMBERSHIP DIRECTORY

Last Name	First Name	Organization	Address	City	State	Zip	Phone No.	Email Address
Abbey	**Tim**	Penn State Cooperative Extension	112 Pleasant Acres Road	York	PA	17402	717-840-7408	tma13@psu.edu
Ackley	**Bruce**	The Ohio State University	2021 Coffey Rd.	Columbus	OH	43210	740-225-2014	ackley.19@osu.edu
Agnew	**Michael**	Syngenta Crop Protection	302 Rose Glen Lane	Kennett Square	PA	19348	610-444-2063	michael.agnew@syngenta.com
Alea	**Stephanie**	Rutgers University	59 Dudley Rd	New Brunswick	NJ	08901	732-932-9711 x 116	stephya@eden.rutgers.edu
Allen	**Jayla**	Bayer Crop Science	Research Triangle Park	Raleigh	NC	27709		jayla.allen@bayer.com
Armel	**Greg**	BASF Corp.	26 Davis Drive	Research Triangle Park	NC	27709	919-547-2674	gregory.armel@basf.com
Arsenovic	**Marija**	Rutgers - IR-4 Headquarters	500 College Ave Suite 201W	Princeton	NJ	08540	732-932-9575 x 4609	arsenovic@aesop.rutgers.edu
Ashley	**James**	Evonik Goldschmidt Corporation	710 South 6th Avenue	Hopewell	VA	23860	804-452-5692	james.ashley@evonik.com
Askew	**Shawn**	Virginia Tech	435 Old Glade Road	Blacksburg	VA	24061	540-231-5807	saskew@vt.edu
Atland	**James**	USDA-ARS ATRU	208 Ag. Eng. Building	Wooster	OH	44691	330-263-3870	James.altland@ars.usda.gov
Averill	**Kristine**	Penn State University	116 ASI Building	University Park	PA	16802	860-248-9969	kma236@psu.edu
Baker	**Robert**	Scotts Company	14111 Scottslawn Rd.	Marysville	OH	43041	937-645-2628	robert.baker@scotts.com
Barcel	**David**	OHP Inc.	W331 S4122 Saddleback Dr.	Genesee	WI	53118	262-392-3004	dbarcel@ohp.com
Barkman	**Gary**	Montgomery Weed Control Inc.	18410 Muncaster Road	Derwood	MD	20855	301-503-6024	gbarkmansr@gmail.com
Barney	**Jacob**	Virginia Tech	435 Old Glade Road (0330	Blacksburg	VA	24061	540-231-6323	jnbarney@vt.edu
Barolli	**Sali**	Imperial Nurseries	176 Lost Acres Rd	Granby	CT	06035	860-653-1509	saizba@yahoo.com
Baron	**Jerry**	Rutgers - IR-4 Headquarters	500 College Road, East Suite 201	Princeton	NJ	08540	732-932-9575 x 4605	jbaron@aesop.rutgers.edu
Barrett	**Mike**	Argyle Country Club	14600 Argyle Club Road	Silver Spring	MD	20906	301-774-5102	turfgod1@CS.com

Bates	**Ryan**	Penn State University	116 ASI Building	University Park	PA	16802	814-863-1014	rtb175@psu.edu
Batts	**Roger**	North Carolina State University - IR-4 Center	Box 7523, NCSU Campus	Raleigh	NC	27606	919-515-1668	roger_batts@ncsu.edu
Baxter	**David**	DuPont Crop Protection	Stine-Haskell 1090 Elkton Road	Newark	DE	19714	302-366-5065	david.a.baxter@usa.dupont.com
Becker	**Chris**	BAAR Scientific LLC	6374 Rte. 89	Romulus	NY	14541	607-342-3610	becker89@fltg.net
Bellinder	**Robin**	Cornell University	164 Plant Science Bldg	Ithaca	NY	14853	607-255-7890	rrb3@cornell.edu
Beste	**Ed**	Univ. Maryland, Salisbury Facility (Emeritus)	27664 Nanticoke Road	Salisbury	MD	21801	410-742-8788	ebeste@umd.edu
Bhowmik	**Prasanta**	University of Massachusetts	Stockbridge Hall Room 10	Amherst	MA	01002-2901	413-545-5223	pbhowmik@pssci.umass.edu
Bonanno	**Richard**	University of Massachusetts	255 Merrimack Street	Methuen	MA	01844	978-361-5650	rbonanno@umext.umass.edu
Borger	**Jeffrey**	Penn State University	116 ASI Building	University Park	PA	16802	814-865-3005	jab267@psu.edu
Bostic	**Heidi**	University of Tennessee	252 Ellington Plant Sciences Bldg.	Knoxville	TN	37996	865-974-7324	hbostic@utk.edu
Bowe	**Steve**	BASF Corp.	26 Davis Drive	Research Triangle Park	NC	27709	919-547-2559	steven.bowe@basf.com
Bowers	**Ian**	Pennsylvania Dept Agriculture	2301 North Cameron Street	Harrisburg	PA	17110	814-323-7182	bowers.ian29@gmail.com
Bradley	**Irene**	NEWSS Standing Orders	3059 Sound Avenue	Riverhead	NY	11901	631-727-3595	it21@cornell.edu
Bravo	**Melissa**	Meadow Lake Farm Consulting Services	P.O Box 447	Lawrenceville	PA	16929	814-574-4067	bravomelissa@yahoo.com
Breeden	**Gregory**	University of Tennessee	252 Ellington Plant Sciences	Knoxville	TN	37996	865-974-7208	gbreeden@utk.edu
Brill	**Nancy**	Syngenta	128 Federal Walk	Kennett Square	PA	19348	919-306-6094	Nancy.Brill@SYNGENTA.COM
Brooks	**Thomas**	Crop Management Strategies	P.O. Box 510	Hereford	PA	18056	610-767-1944	tjb_cms@fast.net
Brosnan	**James**	University of Tennessee	252 Ellington Plant Sciences	Knoxville	TN	37996	865-974-8603	jbrosnan@utk.edu
Buck	**Jenny**	Wildflower Photographer	121 Curtis Point Drive	Mantoloking	NJ	08738	732-892-5507	jebuck_1@yahoo.com
Buck	**Elizabeth**	Cornell University	905 Bradfield Hall	Ithaca	NY	14853	607- 425-3494	emb273@cornell.edu

Burch	**Patrick**	Dow AgroSciences LLC	3425 Elk Creek Road	Christiansburg	VA	24073	540-382-3062	plburch@dow.com
Burnell	**Keith**	Syngenta Crop Protection	956 Willits Road	Ontario	NY	14519	315-209-7580	keith.burnell@syngenta.com
Cain	**Nancy**	Cain Vegetation Inc.	4 Spruce Boulevard	Acton	ON	L7J 2Y2	519-853-3081	cain.vegetation@gmail.com
Calabro	**Jill**	Valent USA Corporation	1305 Colony Dr	Annapolis	MD	21403	443-235-0612	jill.calabro@valent.com
Cardina	John	Ohio State University	1680 Madison Ave.	Wooster	OH	44691	330-263-3644	cardina.2@osu.edu
Case	**Luke**	The Ohio State University	2001Fayette Ct.	Columbus	OH	43210	614-292-0209	case.49@osu.edu
Chamberlin	**Joe**	Valent USA Corporation	2386 Clower St. Ste. E-100B	Snellville	GA	30078	770-985-0303	jcham@valent.com
Chandran	**Rakesh**	West Virginia University	1076 Agricultural Sci PO Box 6108	Morgantown	VA	26506	304-293-6131	rschandran@mail.wvu.edu
Chism	**Bill**	US EPA, OPP	P.O Box 258	Point of Rocks	MD	21777	703-308-8136	chism.bill@epa.gov
Coffman	**Benjamin**		7920 Forest and Stream Club Road	Keymar	MD	21757	443-547-1225	bencoffman41@yahoo.com
Corbett	**Jerry**	Quali-Pro	423 Batten Pond Road	Selma	NC	27576	919-279-2584	jerryc@quali-pro.com
Cox	**Michael**	Virginia Tech	435 Old Glade Road	Blacksburg	VA	24061	540-231-5835	mcc83@vt.edu
Cranmer	**John**	Valent USA Corporation	2228 Glengate Circle	Morrisville	NC	27560	919-387-2013	jcran@valent.com
Crockett	**Benjamin**	Penn State University	420 ASI Building	University Park	PA	16802	814-863-7112	bcc158@psu.edu
Cuddeback	**Richard**	Bluegrass Lawn Service and Landscaping	7201 Hamilton Blvd.	Allentown	PA	18915		
Cummins	**John**	University of Tennessee	1705 Lee Cr.	New Market	TN	37820	865-228-9789	recon77@utk.edu
Curran	**William**	Penn State University	116 ASI Building	University Park	PA	16802	814-863-1014	wcurran@psu.edu
Cushman	**Mark**	The Scott Company	14111 Scottslawn Rd.	Marysville	OH	43041	937-644-0011	mark.cushman@scotts.com
Custis	**Gary**	PBI Gordon Corporation	1217 W. 12th Street	Kansas City	MO	64101	816-460-6215	gcustis@pbigordon.com
Cutulle	**Matthew**	Syngenta	7145 58th Avenue	, Vero Beach	FL	32967	772-794-7170.	Matt.Cutulle@SYNGENTA.COM

D'Appollonio-Cote	**Jennifer**	University of Maine	5722 Deering Hall	Orono	ME	04469	207-581-2924	jennifer.dappollonio@maine.edu
David	**Paul**	Gowan Company	343 Rumford Road	Lititz	PA	17543	717-560-8352	pdavid@gowanco.com
Davis	**James**	Vegetation Managers Inc.	5574 Clearfield Woodland Hwy	Clearfield	PA	16830	814-765-5875	vminc1@verizon.net
Davis	**Todd**	Delaware Dept. Agriculture	2320 S Cedar Crest Blvd	Dover	DE	19901	302-697-4468	Todd.Davis@state.de.us
Davis	**Vince**	University of Illinois	320 ERML, 1201 W. Gregory Dr.	Urbana	IL	61801	217-333-1531	davisv@illinois.edu
Derhoeden	**Peter**	University of Maryland	1112 H.J. Patterson Hall	College Park	MD	20742	301-405-1337	pd@umd.edu
Derr	**Jeffrey**	Virginia Tech	Hampton Rds, AREC 1444 Diamond Spring	Virginia Beach	VA	23455	757-363-3912	jderr@vt.edu
Dillehay	**Bryan**	Monsanto Company	14806 Harrisville Road	Mt. Airy	MD	21771	814-404-2683	bryan.dillehay@monsanto.com
Dillon	**Cory**	Penn State University	116 ASI Building	University Park	PA	16802	814-863-1014	csd109@psu.edu
DiTommaso	**Antonio**	Cornell University	Dept. of Crop & Soil 903 Bradfield Hall	Ithaca	NY	14853	607-254-4702	ad97@cornell.edu
Dobbs	**Jeffrey**	OHP Inc.	1095 Applecross Drive	Roswell	GA	30075	770-992-0121	jdobbs@ohp.com
Dougherty	**Ryan**	Virginia Tech	435 Old Glade Road	Blacksburg	VA	24061	540-231-6323	dougherty@vt.edu
Druist	**Dervin**	Syngenta	18166 Dry Run Road W	Spring Run	PA	17262	717-360-8322	dervin.druist@syngenta.com
Dufoe	**Allan**	FMC Corporation		Spring Run,PA 17262-9720	PA	19106	410-570-6206	allan.d.dufoe@fmc.com
Dunst	**Richard**	Cornell Lake Erie Research and Extension Lab	6592 West Main Rd.	Portland	NY	14769	716-785-9910	rdunst@netsync.net
Dutt	**Timothy**	LABServices	342 South Third Street	Hamburg	PA	19526	610-562-5055	tedutt@ptd.net
Egan	**Franklin**	Penn State University	116 ASI Building	University Park	PA	16802	814-865-6679	jfe121@psu.edu
Ekins	**Rick**	FMC Professional Solutions	1735 Market Street	Philadelphia	PA	19103	215-299-5836	rick.ekins@fmc.com
Elmore	**Matthew**	University of Tennessee	252 Ellington Plant Sciences	Knoxville	TN	37996	865-974-7324	melmore6@utk.edu
Estes	**Tony**	United Phosphorus Inc.	206 Stonewall Heights	Abingdon	VA	24210	864-202-7526	tony.estes@uniphos.com

Evans	**Glenn**	Cornell University	146A Plant Science Bldg.	Ithaca	NY	14853	607-255-9085	gje2@cornell.edu
Everman	**Wesley**	North Carolina State University	7620 Williams Hall	Raleigh	NC	27695	919-518-0488	wes_everman@ncsu.edu
Farrington	**Steven**	Gowan Company	724 Wrights Mill Road	Auburn	Al	36830	334-275-5396	sfarrington@gowanco.com
Fausey	**Jason**	Valent USA Corporation	111 W. Co. Rd. 173	Fremont	OH	43420	419-307-7666	jason.fausey@valent.com
Fidanza	**Michael**	Penn State University	Berks Campus, 2080 Tulpehocken Road	Reading	PA	19610	610-396-6330	maf100@psu.edu
Forney	**Raymond**	Dupont Crop Protection	Stine-Haskell 1090 Elkton Road	Newark	DE	19714	302-561-0027	raymond.forney@usa.dupont.com
Gallandt	**Eric**	University of Maine	7 Mountain View Drive	Orono	ME	04473	207-581-2933	gallandt@maine.edu
Ganske	**Donald**	DuPont Crop Protection	125 Cotton Ridge Rd	Winchester	VA	22603	540-662-6011	donald.d.ganske@usa.dupont.com
Garbett	**Mel**	Central Garden and Pet	1340 Treat Blvd., Suite 600	Walnut Creek	CA	94597	925-948-4000	mgarbett@central.com
Ghantous	**Katherine**	UMass Cranberry Station	PO Box 569, 1 State Bog Rd	East Wareham	MA	02538	508-295-2212	kghantous@psis.umass.edu
Gianessi	**Leonard**	CropLife Foundation	1156 15th Street NW	Washington	DC	20005	202-872-3865	lgianessi@croplifefoundation.org
Gilliam	**Charles**	Auburn University	101 Funchess Hall	Auburn	AL	36849	334-844-3045	murphan@auburn.edu
Glasgow	**Les**	Syngenta	410 S Swing Rd	Greensboro	NC	27409	336-632-5501	les.glasgow@syngenta.com
Glenn	**Scott**	University of Maryland	0115 HJ Patterson Hall	College Park	MD	20742	301-405-1331	sglenn@umd.edu
Goddard	**Matt**	Virginia Tech	435 Old Glade Road	Blacksburg	VA	24061	540-231-5835	mgoddard@vt.edu
Gomez de Barreda	**Diego**	Polytechnic University of Valencia	1109 Experiment Street, Crop and Soil Sci. Dept	Griffin	GA	30223	678-662-5003	diegode@btc.upv.es
Gover	**Art**	Penn State University	116 ASI Building	University Park	PA	16802	814-863-9904	aeg2@psu.edu
Graham	**James**	Monsanto (Retired)	12381 Country Glen Lane	Creve Coeur	MO	63141	314-805-7867	jcgrah@charter.net
Graham	**Ian**	Penn State University	116 ASI Building	University Park	PA	16802	814-863-1014	img102@psu.edu
Graves	**Dean**	Chevy Chase Club	6100 Connecticut Avenue	Chevy Chase	MD	20815	301-656-6323	deang@chevychaseclub.org

Green	**Jerry**	Pioneer	1521 Yeatmans Station Road	Landenberg	PA	19350	484-574-7761	jerry.m.green@pioneer.com
Guiser	**Scott**	Penn State Cooperative Extension	1282 Almshouse Road	Doylestown	PA	18901	215-345-3283	sxg6@psu.edu
Gylling	**Steve**	Gylling Data Management	405 Martin Boulevard	Brookings	SD	04605	605-692-4021	staff@gdmdata.com
Hahn	**Russell**	Cornell University	238A Emerson Hall	Ithaca	NY	14853	607-255-1759	rrh4@cornell.edu
Han	**Kyung**	Penn State University	116 ASI Building	University Park	PA	16802	814-863-1014	kmh373@psu.edu
Hannig	**Gregory**	Dupont Crop Protection	1199 Canandaigua Rd.	Palmyra	NY	14522	302-299-7304	greg.hannig@usa.dupont.com
Hanrahan	**Richard**	Bayer Environmental Science	100 East Paliside Ave C-42	Englewood	NJ	07631	201-294-5217	rich.hanrahan@bayercropscience.com
Harlow	**Chris**	North Carolina State University	70 Kilgor Hall, Box 7609	Raleigh	NC	27695	919-515-8097	chris_harlow@ncsu.edu
Harpster	**Tracy**	Penn State University	102 Tyson Building	University Park	PA	16802	814-865-3190	tlh8@psu.edu
Hart	**Stephen**	Rutgers University	59 Dudley Rd	New Brunswick	NJ	08901	732-932-9711 x 166	hart@aesop.rutgers.edu
Hearn	**Brian**	University of Delaware	16483 County Seat Highway	Georgetown	DE	19947	302-856-7303	bhearn@udel.edu
Hedberg	**Rob**	USDA National Institute of Agriculture	5612 Mclean Drive	Bethesda	MD	20814	202-720-5384	rhedberg@nifa.usda.gov
Heimer	**Lane**	Maryland Department of Agriculture	100 West Washington Street	Hagertown	MD	21740	410-841-5920	laneheimer@gmail.com
Herrick	**Robert**	FMC Corporation	11 Wolfpack Court	Hamilton	NJ	08619	609-951-3792	bob.herrick@fmc.com
Hess	**Dwayne**	J.C. Ehrlich Co., Inc.	500 Spring Rdige Drive P.O. Box 13848	Reading	PA	19612	610-372-9750	dwayne.hess@jcehrlich.com
Hester	**Kathleen**	Temple University	508 Heather Rd	Exton	PA	19341	484-868-2063	hester@aesop.rutgers.edu
Higgins	**Simone**	Monsanto	800 N. Lindbergh Blvd	St. Louis	MO	63167	314-330-3053	simone.seifert-higgins@monsanto.com
Hines	**Thomas**	Eastern Shore AREC VPI/SU	33446 Research Drive	Painter	VA	23420	757-414-0724	thhines@vt.edu
Hitchner	**Erin**	Syngenta Crop Protection	380 Jefferson Rd	Elmer	NJ	08318	609-980-8832	erin.hitchner@syngenta.com
Hodges	**Duane**	Scotts Company	14111 Scottslawn Rd.	Marysville	OH	43041	email him	duane.hodges@scotts.com

Hoffer	**Michael**	BASF Corp.	26 Davis Drive	Research Triangle Park	NC	27709	919-244-8973	mike.hofer@basf.com
Hooten	**Robert**	FMC Corporation	10100 N. Ambass. Drive Suite 400	Liberty	MO	60468	816-589-5547	robert_hooten@fmc.com
Horton	**Chris**	North Carolina State University	Campus Box 7620	Raleigh	NC	27695	919-515-7597	chris_reberg-horton@ncsu.edu
Houseworth	**Doug**	Arysta LifeScience	2777 Ocean Oaks Dr. S.	Fernandina Beach	FL	32034	904-321-0795	doug.houseworth@arystalifescience.c om
Huffman	**Janel**	Penn State University	116 ASI Building	University Park	PA	16802	814-863-1014	janelhuffman1988@gmail.com
Hutto	**Kendall**	FMC Professional Solutions	136 Spring Valley Rd.	Westerville	OH	43081	614-392-1384	kendall.hutto@fmc.com
Ikley	**Joseph**	University of Maryland	6423 Church Street	Sykesville	MD	21784	410-596-9091	jikley@umd.edu
Isaacs	**Mark**	University of Delaware	16483 County Seat Highway	Georgetown	DE	19947	302-856-7303	isaacs@udel.edu
James	**J.R.**	Syngenta Crop Protection	410 Swing Road	Greensboro	NC	27409	336-632-6000	j_r.james@syngenta.com
Jemison	**John**	University of Maine - Cooperative Extension	495 College Avenue	Orono	ME	04469	207-581-3241	jjemison@umext.maine.edu
Jester	**Jen**	Virginia Tech	435 Old Glade Road	Blacksburg	VA	24061	540-231-5835	jesterj@vt.edu
Johnson	**Jon**	Penn State University	102 Tyson Bldg.	University Park	PA	16802	814-863-1184	jmj5@psu.edu
Johnson	**Quintin**	University of Delaware	Carvel REC 16483 County Seat Hwy	Georgetown	DE	19947	302-856-2585	quintin@udel.edu
Johnson	**David**	Pioneer Research	7250 NW 62nd Ave	Johnston	IA	50131	515-727-7234	david.h.johnson@pioneer.com
Johnson	**Roy**	Waldrum Specialties Inc	1727 E Butler Pike	Ambler	PA	19002-2431	215-817-0637	rjoh834880@aol.com
Jones	**Patrick**	University of Tennessee	2431 Joe Johnson Drive	Knoxville	TN	37996	865-974-6730	patajone@utk.edu
Jordan	**Grant**	A.C.D.S. Research Inc.	9813 Glenmark Road	North Rose	NY	14516	315-587-2140	gljordan@acdsresearch.com
Jotcham	**Jim**	Marbicon Inc.	Box 280, Berwick	Nova Scotia	Canada	B0P 1E0	902-538-7101	marbicon@eastlink.ca
Kahl	**Jerry**	J.C. Ehrlich Co., Inc.	500 Spring Ridge Drive P.O. Box 13848	Reading	PA	19612	610-372-9750	jerry.kahl@jcehrlich.com
Kalmowitz	**Kathie**	BASF Corp.	3955 Stags Leap Circle	Raleigh	NC	27612	919-270-4592	kathie.kalmowitz@basf.com

Kaminski	**John**	Penn State University	442 Ag Admin. Building	University Park	PA	16802	814-865-3007	kaminksi@psu.edu
Kammerer	**Melanie**	Penn State University	116 ASI Building	University Park	PA	16802	814-863-1014	
Kao-Kniffin	**Jenny**	Cornell University	419 Mitchell St.	Ithaca	NY	14850	607-255-8886	jtk57@cornell.edu
Keene	**Clair**	Penn State University	116 ASI Building	University Park	PA	16802	814-863-1014	clk5285@psu.edu
Keese	**Renee**	BASF Corp.	26 Davis Dr	Research Triangle Park	NC	27709	919-547-2791	renee.keese@basf.com
Kelly	**Patrick**	Anne Arundel County Recreation and Parks Department	301 Hope Road	Centreville	MD	21617	443-262-9655	rpkell96@aacounty.org
Kirfman	**Gary**	Valent USA Corporation	6088 Winthrop Ave.	Ada	MI	49301	616-560-2510	gkirf@valent.com
Kleczewski	**Victoria**	DuPont	463 W Harvest Ln	Middletown	DE	19709	219-208-9499	victoria.a.kleczewski@dupont.com
Knight	**Alexandra**	North Carolina State University	Box 7620 Williams Hall	Raleigh	NC	27695		amknigh4@ncsu.edu
Koo	**Suk-Jin**	Moghu Research Center		Daejeon		South Korea		moghu@moghu.com
Kordbacheh	**Farnaz**	Cornell University	903 Bradfield Hall Dept. of Crop & Soil	Ithaca	NY	14850	410-830-0781	fk226@cornell.edu
Kunkel	**Dan**	Rutgers - IR-4 Headquarters	500 College Ave Suite 201W	Princeton	NJ	08540	732-932-9575	kunkel@aesop.rutgers.edu
Kyde	**Kerrie**	Maryland Department of Natural Resources	WHS 11960 Clopper Road	Gaithersburg	MD	20878	301-948-8243	kkyde@dnr.state.md.us
Larson	**Steven**	BASF Corp.	26 Davis Dr.	Research Triangle Park	NC	27709	281-684-4542	steven.larson@basf.com
Leahey	**Caitlin**	Cornell University	503 East Buffalo St, Apt 3	Ithaca	NY	14853	607-255-4747	caitlin.leahey@gmail.com
Leather	**Gerald**	West Virginia University Extension Service	66 N. High St., P.O. Box 1880	Romney	WV	26757	304-822-5013	gerald.leather@mail.wvu.edu
Lenker	**Jessica**	Pennsylvania Department of Agriculture | Bureau of Plant Industry	2301 North Cameron Street	Harrisburg	PA	01710	717-787-7204	jeslenker@pa.gov
Lightfoot	**Rachel**	Crop Management Strategies	P.O. Box 510	Hereford	PA	18056	610-767-1944	ral_cms@fast.net
Linde	**Douglas**	Delaware Valley College	700 E. Butler Ave.	Doylestown	PA	18901	215-489-2260	douglas.linde@delval.edu

Lindquist	**John**	University of NebraskA	105CKCR	Lincoln	NE	68583-0817	402-472-2771	jlindquist1@unl.edu
Lingenfelter	**Dwight**	Penn State University	116 ASI Building	University Park	PA	16802	814-865-2242	dwight@psu.edu
Little	**Neith**	Cornell University	Room 907 Bradfield Hall	Ithaca	NY	14853	607-346-5958	ngl7@cornell.edu
Little	**Daniel**	North Carolina State University	5616B Thea Ln	Raleigh	NC	27606	517-974-3000	dalittle@ncsu.edu
Lloyd	**Kirsten**	Penn State University	102 Tyson Bldg	University Park	PA	16802	814-863-1184	kll24@psu.edu
Loecke	**David**	PBI Gordon	5701 North Chatham Ct.	Kansas City	MO	64151	816-547-8373	vcline@pbigordon.com
Lohmann	**Henry**	Nicholls University (Retired)	P.O. Box 22	Bellport	NY	11713	631-286-1078	halohmann@aol.com
Lough	**Katherine**	University of Maine	5722 Deering Hall	Orono	ME	04469	207-581-2923	klough@maine.edu
Loughner	**Dan**	Dow AgroSciences LLC	55 Armour Road	Princeton	NJ	08540	609-921-1856	dloughner@Dow.com
Lurvey	**Edith**	Cornell University - NYSAES	630 W. North St.	Geneva	NY	14456	315-787-2308	ell10@cornell.edu
Lycan	**Darren**	Syngenta Lawn and Garden	18 Appleridge St.	Baldwinsville	NY	13027	315-635-2818	darren.lycan@syngenta.com
Lydon	**John**	USDA-ARS	5601 Sunnyside Ave GWCC Rm 4-2238	Beltsville	MD	20705-5139	301-504-6470	john.lydon@ars.usda.gov
Lyman	**Betsy**	National Park Service	1 River Rd	Bushkill	PA	18324	570-588-0513	betsy_lyman@nps.gov
Mahoney	**Matthew**	Bayer CropScience	4773 Sailors Retreat Rd	Oxford	MD	21654	410-822-5215	matt.mahoney@bayer.com
Majek	**Bradley**	Rutgers University	1128 Rainbow Circle	Pittsgrove	NJ	08318	856-455-3100	majek@aesop.rutgers.edu
Manley	**Brian**	Syngenta Biotechnology, Inc.	3054 E. Cornwallis Road	Durham	NC	27709	919-281-2162	brian.manley@syngenta.com
Mansue	**Carrie**	Rutgers University	59 Dudley Rd	New Brunswick	NJ	08901	732-932-9711 x 116	cmansue@aesop.rutgers.edu
Marble	**Chris**	Auburn University	101 Funchess Hall	Auburn	AL	36849		marblsc@auburn.edu
Marose	**Betty**	USDA-ARS	10300 Baltimore Ave Building 001 Room 110 BARC-West	Beltsville	MD	20705	301-504-5156	bhmarose@comcast.net
Marschner	**Caroline**	Cornell University	90 Penny Lane	Ithaca	NY	14850	360-915-4778	cam369@cornell.edu

Mathers	**Hannah**	Ohio State University	256 B Howlett Hall	Columbus	OH	43210	614-247-6195	mathers.7@osu.edu
Mathews	**Sudeep**	University of Maryland	501 Court Lane P. O. Box 299	Cambridge	MD	21613	410-228-8800	samathew@umd.edu
Matthews	**Michael**	For-Shore Weed Control	P.O. Box 536	Waretown	NJ	08757	609-693-6999	mike@for-shore.com
Mayonado	**David**	Monsanto Company	6075 Westbrooke Drive	Salisbury	MD	21801	410-726-4222	david.j.mayonado@monsanto.com
McCullough	**Patrick**	University of Georgia	1109 Experiment St	Griffin	GA	30223	770-228-7276	pmccull@uga.edu
McDonald	**Steven**	Turfgrass Disease Solutions LLC	38 Bertolet School Road	Spring City	PA	19475	610-633-1878	turfgrassdiseasesolutions@yahoo.com
McDonnell	**Brian**	National Park Service	1 River Rd	Bushkill	PA	18324	570-588-0534	brian_mcdonnell@nps.gov
McNulty	**Brendon**	Virginia Tech	435 Old Glade Road	Blacksburg	VA	24060	540-231-5807	bmcnulty@vt.edu
Meck	**Elijah**	Syngenta	7145 58th Avenue	Vero Beach	FL	32967	1 772 5675218	elijah.meck@syngenta.com
Mervosh	**Todd L.**	Connecticut Ag. Exp Station	153 Cook Hill Road	Windsor	CT	06095	860-683-4984	todd.mervosh@ct.gov
Meyer	**Tanya**	Sinn. Invasive Plant Mngt. Area	PO Box 123	Emporium	PA	15834	814-486-1740	tanya@sipmapa.org
Milbrath	**Lindsey**	USDA-ARS Robert W. Holley Center	538 Tower Road	Ithaca	NY	14853	607-254-7268	lindsey.milbrath@ars.usda.gov
Miller	**Raymond**	Dow AgroSciences LLC	4020 Collinwood Ave	Fort Worth	TX	76107	813-363-9059	rcmiller2@dow.com
Mirsky	**Steven**	USDA-ARS ANRI SASL	10300 Baltimore Ave Building 001 Room 110 BARC-West	Beltsville	MD	20705	301-504-7199	steven.mirsky@ars.usda.gov
Mohler	**Charles**	Cornell University	907 Bradfield Hall	Ithaca	NY	14853	607-255-0199	clm11@cornell.edu
Morris	**Tom**	University of Connecticut-SARE	1376 Storrs Road	Storrs	CT	06269	860-486-0637	thomas.morris@uconn.edu
Mortensen	**David**	Penn State University	116 ASI Building	University Park	PA	16802	814-865-6679	dmortensen@psu.edu
Mosdell	**Dean**	Syngenta Crop Protection	501-I S. Reino Rd. #183	Newbury Park	CA	91320	805-480-0514	dean.mosdell@syngenta.com
Munsterman	**Chris**	Syngenta Crop Protection	11 Quicksilver Court	Martinsburg	WV	25404	304-261-9564	chris.munsterman@syngenta.com
Myers	**Donald D.**	Bayer Crop Science	2 T.W. Alexander Drive	Research Triangle Park	NC	27709	919-549 2529	don.myers@bayer.com

Naedel	**Matthew**	Penn State University	Valentine Turfgrass Ctr. Univ. Drive Ext.	University Park	PA	16802	814-863-1613	mbn112@psu.edu
Neal	**Joseph**	North Carolina State University	262 Kilgore Hall Box 7609	Raleigh	NC	27606	919-515-9379	joe_neal@ncsu.edu
Norton	**Lawrence**	Bayer Environmental Science	4233 Harriet Lane	Bethlehem	PA	18017	610-814-6220	larry.norton @bayercropscience.com
O'Barr	**John H.**	BASF Corp.	108 Whippoorwill Lane	Hummelstown	PA	17036	717-386-8259	john.obarr@basf.com
O'Connell	**James**	UMass Cranberry Station	P.O Box 569	East Wareham	MA	02538	508-295-2212	jameso@umext.umass.edu
Olson	**Brian**	Dow AgroSciences LLC	22 Delancey Dr.	Geneva	NY	14456	317-337-7245	bdolson@dow.com
O'Neal	**William**	AMVAC Chemical Corporation	102 Bay View Drive	Chapel Hill	NC	27516	919-619-3095	boneal@sprynet.com
O'Neill	**Brian**	Weeds Inc	250 Bodley Rd	Aston	PA	19014	610-358-9430	bgon58@aol.com
O'Neill	**Drew**	Weeds Inc	250 Bodley Rd	Aston	PA	19014	610-358-9430	weeds@weedsinc.com
Orlowski	**John**	University of Kentucky	1405 Veterans Drive	Lexington	KY	40546	518-596-9311	john.orlowski@uky.edu
Orr	**James**	Asplundh	708 Blair Mill Road	Willow Grove	PA	19090	800-248-TREE	jimorr@asplundh.com
Pacchioli	**Marc**	Crop Management Strategies	P.O. Box 510	Hereford	PA	18056	610-767-1944	cms1@fast.net
Palmer	**Cristi**	Rutgers - IR-4 Headquarters	500 College Ave Suite 201W	Princeton	NJ	08540	732-932-9575	palmer@aesop.rutgers.edu
Palmer	**Butch**	Reality Research	5916 South Ave.	Williamson	NY	14589	315-945-0945	whpalmer@localnet.com
Pannill	**Philip**	U.S. Fish and Wildlife	698 Conservation Way	Shepherdstown	WV	25443	304-876-7432	phil_pannill@fws.gov
Parker	**Astrid**	Bayer Environmental Science	981 NC42 East	Clayton	NC	27527	919-625-3350	astrid.parker@bayer.com
Parochetti	**James**	retired and a distinguished member	3713 Whispering Lane	Falls Church,	VA	22041	202-401-4354	jamesparochetti@aol.com
Parrish	**Jason**	Ohio State University	2001 Fyffe Ct 256 Howlett Hall	Columbus	OH	43210	440-225-4831	parrish.174@osu.edu
Patches	**Kelly**	Penn State University	116 ASI Building	University Park	PA	16802	814-863-1014	kmp5212@psu.edu
Patton	**Aaron**	Purdue University	Lilly Hall 915 West State Street	West Lafayette	IN	47907-2054	765-494-9737	ajpatton@purdue.edu

Pawlak	**John**	Valent USA Corporation	1333 N California Blvd Ste 600	Walnut Creek	CA	look up	925-256-2700	jpawlak@valent.com
Pennucci	**Annamarie**	Northeast Turf & Ornamental Research	4 Englewood Drive	Raymond	NH	03077	610-895-8480	aapennucci@yahoo.com
Phillips	**Jeffery**	Monsanto	42 Botsford Street	Poland	OH	44514	330-402-2591	jeffrey.i.phillips@monsanto.com
Phillips	**Bill**	US EPA, OPP	1200 Pennylvania Ave., N.W.	Washington	DC	20460	703-308-8099	phillips.bill@epa.gov
Pieczarka	**David**	Gowan Company	1630 Berry Road	Lafayette	NY	13084		Pieczarka@gowanco.com
Poffenbarger	**Hannah**	USDA-ARS ANRI SASL	10300 Baltimore Ave Building 001 Room 110 BARC-West	Beltsville	MD	20705	301-504-7199	hpoffenb@gmail.com
Post	**Angela**	Virginia Tech	435 Old Glade Road Campus Box 0330	Blacksburg	VA	24060	540-231-5807	arpost@vt.edu
Prostak	**Randy**	University of Massachusetts	Dept. Plant & Soil French Hall Room 2	Amherst	MA	01033	413-577-1738	rprostak@umext.umass.edu
Pruss	**Stanley**	CIBA-Geigy (Retired)	443 Moninger Rd	Washington	PA	15301	724-222 4831	spruss@hughes.net
Pyle	**Steve**	Syngenta Crop Protection	410 Swing Road	Greensboro	NC	27409	336-632-2236	steve.pyle@syngenta.com
Qin	**Zhong**	South China Agriculture University	c/o Cornell University 905 Bradfield Hall	Ithaca	NY	14853	607-254-5365	q_breeze@126.com
Rana	**Aman**	Virginia Tech	813 Flaton Trace	Chesapeake	VA	23322	757-237-3412	ranaaman29@gmail.com
Rardon	**Patrick**	Dupont Crop Protection	Stine-Haskell 1090 Elkton Road	Newark	DE	19711	302-366-5546	patrick.l.rardon@usa.dupont.com
Ratnayake	**Sunil**	US EPA	1200 Pennsylvania Ave. N.W. Mail Stop 7503 P	Washington	DC	20460	703-308-8191	ratnayake.sunil@epa.gov
Reeb	**Bryan**	The Ohio State University	2021 Coffey Rd.	Columbus	OH	43210	740-225-2014	reeb.22@osu.edu
Reed	**Joseph**	FMC Corportion	5284 Mayberry Pl.,	Dayton	OH	45415	217-649-3249	joseph.reed@fmc.com
Richardson	**Robert**	North Carolina State University	Box 7620 Williams Hall	Raleigh	NC	27695	919-515-5653	rob_richardson@ncsu.edu
Richtmyer III	**Richard**	Cornell University	3179 County Line Road	Watkins Glen	NY	14891	607-342-1695	rjr39@cornell.edu
Riffle	**Michael**	Valent USA Corporation	P.O Box 8025	Walnut Creek	CA	94596		mriff@valent.com

Ritter	**Ronald**	University of Maryland	12901 North Point Lane	Laurel	MD	20708	301-405-1329	rlritter@umd.edu
Rogers	**Gregory**	DuPont Crop Protection	58 Middlecroft Rd	Elkton	MD	21921	443-309-0148	gregory.s.rogers@usa.dupont.com
Rorem	**Kent**	North Carolina Department of Agriculture	3800 Castle Hayne Rd.	Castle Hayne	NC	28429	910-675-2314	kent.rorem@ncagr.gov
Roten	**Rory**	North Carolina State University	Campus Box 7620	Raleigh	NC	27695	919-610-6165	rory_roten@ncsu.edu
Rupp	**Peter**	Maryland Department of Agriculture	50 Harry S. Truman Pkwy.	Annapolis	MD	21401	410-841-5920	halletdc@mda.state.md.us
Ryan	**Matthew**	Cornell University	116 ASI Building	University Park	PA	16802	814-574-3028	mrr203@psu.edu
Sandler	**Hilary**	UMass Cranberry Station	P.O. Box 569	East Wareham	MA	02538	508-295-2212 x 21	hsandler@umext.umass.edu
Sandy	**Dave**	Penn State University	116 ASI Building	University Park	PA	16802	814-404-9001	djs230@psu.edu
Saunders	**David**	DuPont Crop Protection	24087 230th st.	Dallas Center	IA	50063	515-334-4485	david.w.saunders@usa.dupont.com
Schou	**Bert**	ACRES Research	P.O. Box 99	Green Bank	WV	24944	304-456-5558	bertschou@aol.com
Scoresby	**Rene**	Green Light	1058 Blueberry Lane	Mosinee	WI	54455	715-298-3315	renescoresby@yahoo.com
Scott	**Barbara**	University of Delaware	16483 County Seat Hwy	Georgetown	DE	19947	302-856-2585	bascott@udel.edu
Sellman	**Leroy**	Maryland Department of Agriculture	50 Harry S. Truman Pkwy.	Annapolis	MD	21401	410-841-5920	halletdc@mda.state.md.us
Senesac	**Andrew**	Cornell University Cooperative Ext.	3059 Sound Avenue	Riverhead	NY	11901	631-727-3595	afs2@cornell.edu
Serensits	**Thomas**	Penn State University	116 ASI Building	University Park	PA	16802	610-360-5985	tjs204@psu.edu
Sharma	**Shiv**	FMC Corporation	1735 Market Street	Philadelphia	PA	19103	215-299-6871	shiv.sharma@fmc.com
Shirtliffe	**Steve**	University of Saskatchewan	3D04 - 51 Campus Drive	Saskatoon	SK	s7H279 Canada	306-966-4959	steve.shirtliffe@usask.ca
Smith	**Richard**	University of New Hampshire	56 College Rd	Durham	NH	03824	603-862-2724	richard.smith@unh.edu
Smith	**Mark**	Maryland Department of Agriculture	50 Harry S. Truman Pkwy.	Annapolis	MD	21401	410-841-5920	smithmj@mda.state.md.us
Smith	**Adam**	Virginia Tech	435 Old Glade Road	Blacksburg	VA	24061	540-231-5835	urzsmith@vt.edu

Smith	**Larissa**	Virginia Tech	435 Old Glade Road	Blacksburg	VA	24061	540-231-6323	lls14@vt.edu
Smith	**Chris**	Bluegrass Lawn Service and Landscaping	PO Box 385	Southampton	PA	18966	215-364-7000	Service@BGLawnService.com
Snyder	**Eli**	Penn State University	116 ASI Building	University Park	PA	16802	814-863-1014	elina.snyder@gmail.com
Spak	**David**	Bayer Environmental Science	981 Highway 42 East	Clayton	NC	27520	919-625-3350	david.spak@bayer.com
Stachowski	**Paul**	Cornell University	1256 Poplar Ridge Rd.	Aurora	NY	13026	315-364-8296	pjs16@cornell.edu
Stalter	**Richard**	St. John's University	Dept. of Biology 8000 Utopia Parkway	Queens	NY	11439	718-990-6288	stalterr@st.johns.edu
Steffel	**James**	LABServices	342 South Third Street	Hamburg	PA	19526	610-562-5055	jim@labservices.com
Swanton	**Clarence**	University of Guelph	50 Stone Rd. E., Guelph	Ontario	Canada	N1G 2W1	519-824-4120 x 53392	cswanton@uoguelph.ca
Sweet	**Robert**	Cornell University (Emeritus)	c/o Maxine Welcome Dept. Horticulture 167 Plant Sciences Bld	Ithaca	NY	14853	607-273-7106	mw45@cornell.edu
Szylvian	**Andrea**	US EPA, Region 1 Pesticide Program	Post Office Sq. Suite 100 OES 05-4	Boston	MA	02109	617-918-1198	szylvian.andrea@epa.gov
Tan	**Siyuan**	BASF Corp.	26 Davis Dr.	Research Triangle Park	NC	27709	919-547-2679	siyuan.tan@basf.com
Tardif	**Francois**	University of Guelph	50 Stone Rd. E., Guelph	Ontario	Canada	N1G 2W1	519-824-4120 x 53395	ftardif@uoguelph.ca
Taylor	**Erin**	Michigan State University	A285 PSSB, Dept. CSS	East Lansing	MI	48824	517-355-0271	hiller12@msu.edu
Taylorson	**Raymond**	USDA & URI (Retired)	8721 Orchard Dr.	Chestertown	MD	21620	410- 778-9088	raymondtaylorson@verizon.net
Teasdale	**John**	USDA-ARS	10300 Baltimore Ave Building 001 Room 110 BARC-West	Beltsville	MD	20705	301-504-5504	john.teasdale@ars.usda.gov
Tekiela	**Dan**	Virginia Tech	435 Old Glade Road	Blacksburg	VA	24061	540-231-6323	tekiela2@vt.edu
Thomas	**Gar**	BASF Corp.	1002 Bethel Road	Chesapeake City	MD	21915	410-885-5920	garfield.thomas@basf.com
Thomas	**W. Joseph**	BASF Corp.	26 Davis Drive	Research Triangle Park	NC	27709	919-547-2549	walter.e.thomas@basf.com
Toressen	**Kristine**	Cornell University	Room 907 Bradfield Hall	Ithaca	NY	14853	518-596-9311	kst49@cornell.edu

Tranel	**Patrick**	University of Illinois	320 ERML 1201 W. Gregory Dr.	Urbana	IL	61801	217-333-1531	tranel@illinois.edu
True	**Sarah**	North Carolina State University	Campus Box 7620	Raleigh	NC	27695	919-812-5679	sltrue@ncsu.edu
Vail	**Gordon**	Syngenta Crop Protection	410 Swing Road	Greensboro	NC	27409	336-632-5596	gordon.vail@syngenta.com
Vanderwerker	**Cindy**	National Park Service	648 Rt. 32	Stillwater	NY	12170	518-664-9821	cindy_vanderwerker@nps.gov
VanGessel	**Mark**	University of Delaware	16483 County Seat Hwy	Georgetown	DE	19947	302-856-7303	mjv@udel.edu
Vanini	**J.Tim**	New Dimensions Turf	9 Colvin Ave	Buffalo	NY	14216	716-319-7495	tim@ndturf.com
Vargas	**J.Javier**	University of Tennessee	252 Ellington Plant Sci. Bldg.	Knoxville	TN	37996	865-974-7324	jvargas@utk.edu
Vea	**Elymar V.**	IR-4 Headquarters	308 Aston Forest Lane	Crownsville	MD	21032	410-923-4880	evvea@comcast.net
Venner	**Katelyn**	Rutgers University	59 Dudley Road	New Brunswick	NJ	08901	732-932-9711 x 116	katevenn@eden.rutgers.edu
Vitolo	**David**	Syngenta Crop Protection	303 W Farriss Ave	High Point	NC	27262	916-316-6951	david.vitolo@syngenta.com
Walls	**Bobby**	FMC Corporation	501 Parkwood Lane	Goldsboro	NC	27530	919-735-3862	bobby.walls@fmc.com
Wells	**M. Scott**	North Carolina State University	Box 7620 NCSU Campus	Raleigh	NC	27695	919-741-9876	mswells@ncsu.edu
Welterlen	**Mark**	PBI/Gordon Corporation	1217 W. 12th Street	Kansas City	MO	64101	816-460-6205	mwelterlen@pbigordon.com
Wenskus	**Tim**	City of NY Parks and Recreation	The Arsenal, Central Park 830 Central Ave	NY	NY	10065	212-New YorK	
West	**Amanda**	North Carolina State University	Campus Box 7620	Raleigh	NC	27695	919-812-5679	amwest@ncsu.edu
White	**Linda**	National Park Service	648 Rt. 32	Stillwater	NY	12170	518-664-9821	linda_white@nps.gov
White	**Tim**	Crop Management Strategies	P.O. Box 510	Hereford	PA	18056	610-767-1944	cms1@fast.net
Whitehouse	**Stephanie**	Cornell University	905 Bradfield Hall	Ithaca	NY	14853	607-255-4747	sew54@cornell.edu
Willis	**John**	Monsanto Company	1305 Sanders Road	Troy	OH	45373	937-418-5667	john.b.willis@monsanto.com
Wilson	**Samuel**	FMC Corportion	113 Arlington Ridge Road	Cary	NC	27513	803-413-7824	Sam.Wilson@fmc.com

Wilson	**Henry**	Virginia Tech	Eastern Shore ARC 33446 Research Drive	Painter	VA	23420	757-414-0724	hwilson@vt.edu
Wollam	**John**	Bayer Crop Science	P.O Box 4913 8400 Hawthorn Road	Kansas City	MO	64120	816 242-2000	john.wollam@bayer.com
Wychen	**Lee**	WSSA Liason	5720 Glen Mullen Place	Alexandria	VA	22303	202-746-4686	Lee.vanWychen@wssa.net
Wynn	**John**	Dupont Crop Protection	Stine-Haskell 1090 Elkton Road	Newark	DE	19714	302-451-5835	w-wynn.john@usa.dupont.com
Yarborough	**David**	University of Maine	5722 Deering Hall	Orono	ME	04469	207-581-2923	davidy@maine.edu
Zawierucha	**Joe**	BASF Corp.	26 Davis Drive	Research Triangle Park	NC	27709	919-547-2095	joseph.zawierucha@basf.com
Zelna	**Jeffrey**	Syngenta Crop Protection	4598 Reliant Rd	Jamesville	NY	13078	315-498-4259	jeff.zelna@syngenta.com
Zhang	**Jia-en**	South China Agricultural University		Guangzhou		Peoples Republic		jeanzh@scau.edu.cn
Zontek	**Stanley**	USGA Green Section, Mid-Atlantic Region	485 Baltimore Pike Suite 203	Glen Mills	PA	19342	610-558-9066	szontek@usga.org

AUTHOR INDEX
(by Abstract Number)

KEYWORD INDEX
(by Abstract Number)